Old Whaling Days

Old Whaling Days

Captain William Barron
MEMBER OF THE BOARD OF THE
HULL TRINITY HOUSE

CONWAY CLASSICS

Dedicated to the wardens and brethren
of the Trinity House, Hull

First published in 1895

First published by Conway Maritime Press in Great Britain in 1970

This edition published in 1996 by Conway Maritime Press in the
Conway Classics series

Conway Maritime Press is an imprint of Brassey's (UK) Ltd,
33 John Street, London WC1N 2AT

© Brassey's (UK) Ltd 1996

British Library Cataloguing in Publication Data

Barron, William
Old Whaling days. - 3rd ed.
1.Barron, William 2.Truelove (Ship) 3.Whaling - Arctic
regions - History - 19th century 4.Whalers (Persons) -
Great Britain - Biography 5.Whaling masters - Great Britain
I.Title
639.2'8'092

ISBN 0-85177-695-7

Printed in Great Britain by Redwood Books, Trowbridge

PREFACE.

THE pages of this book contain a personal narrative of varied experiences in the Northern regions over forty years ago. In the period dealt with, Hull and other ports were largely benefitted by the ships which yearly left the shores of this country for adventurous voyages in the Arctic Seas, and fortunes which many enjoy to-day were laid by ancestors who did not hesitate to brave the perilous and dangerous hardships frequently encountered among the eternal icefields, blinding snowstorms, and fierce gales of the pitiless North. Not a few of England's sons found lonely graves in the land of the Esquimaux and the Polar Bear, and to this day may be found wooden monuments denoting their birth-place, date of death, and the ships to which they belonged. Such were raised by those who reverentially laid them to rest, and as this narrative will show, were found undisturbed in years after. Others who survived to narrate to their children and friends at home thrilling scenes in their Arctic life, and who are alluded to in these pages, have long since embarked on their last voyage to the Silent Land. It is years since the last ship left Hull for Davis's Straits, and the younger generation of Hull's sons know nothing of whaling, and the many essentials necessary for the complete equipment of an Arctic whaler. There are words, too, used in the times dealt with, which do not find a place in the English dictionary of to-day, and it is hoped, therefore, that the information which is given in these pages, will be found full of interest, not only to the uninitiated, but to those who may remember their own experiences in whaling. Thankful for my own escapes in early life, I would beg to lay this narrative before the general public.

THE AUTHOR.

CONTENTS.

OLD WHALING DAYS.

---------◆---------

CHAPTER I.

ENTERING LIFE—BOUND FOR DAVIS'S STRAITS—THE FIRST
WHALE—RETURNING HOME—NEWS OF CHOLERA AT HULL.

ON the 25th March, 1849, I left the Trinity House
School, Hull, having completed three years in that
excellent institution for training boys for the sea service.

On the 27th of the same month, I was bound apprentice
in the barque "Truelove," 296 tons register, which was built
in Philadelphia in 1764. She had carried a letter of marque,
and fought her own convoy during the Napoleonic wars.
Her owner was T. Ward, Esq., at that time one of the
principal shipowners of Hull. Two days after I was bound
apprentice we sailed for Davis's Straits. Captain J. Parker
was the commander of our ship at this period, and he was
one of the cleverest and most experienced navigators then
sailing to that country. I was to receive £35 for my six
years' service, and an additional remuneration of 6s. per
week during my stay in port. The £35 made an average
of nearly six pounds per year, with which to provide for
clothes and other necessaries. It went a very little way in
those days, especially for such a cold quarter of the world.
I first took up duty as cabin boy, having the master, mate,
and doctor to attend upon, and make myself generally
useful. I had no idle moments. When I went on board, I
was ordered to scrub part of the 'tween decks at a moment's

notice. The crew were absent, excepting the mate, who conceived a dislike for me as soon as I presented myself. His son was with him, and was about my own age. I had not given any provocation, but he was ordered by his father to kick me whilst I was scrubbing the deck. He obeyed the order, and I was very much surprised, not unnaturally, at such cruel treatment. His father told him to do it again, and he repeated the assault. My temper would not stand ill-usage of so cowardly a character, and I got up and knocked the lad down. His father brutally shook me, and vowed he would make me smart for it before I got back to Hull. I subsequently learnt that he had applied for my berth for his son, but the captain had declined to have him. When I was left alone I sat down and wept bitterly, and almost regretted leaving a kind and indulgent mother for a life which promised such hardships at its outset. I did not complain to her, for it had troubled her very much when I first made up my mind to go to sea.

The next day we sailed from Hull, and the cheers of those on shore found a responsive echo in the hearts of our crew. Favoured with a fair wind we proceeded down the Humber to sea. After passing Spurn I began to feel the symptoms of sea-sickness for the first time. The crew were too busy to pay any attention to me, so I sat upon a spare spar feeling far from well. Our captain told me to inform him when I felt sick, and I expected he would have given me some relief. The boatswain was ordered to bring a rope's end, and I then guessed what kind of relief was meant. Captain Parker was too kind, however, to prescribe such a panacea for my indisposition, and I found out that he was joking. He good-humouredly told me to knock about and I should soon be better. Happily bed-time came soon after, eight bells being struck. The watch was set, and I turned into a strange bed for the first time in my life. I shall never forget the feelings I experienced. There

was a voyage of eight months to look forward to, and I felt
very great pain in parting from my friends for that long
time. Amid these melancholy reflections I fell asleep, and
did not wake again until I was called up at six the next
morning by the second mate, a good but rough old man,
and set to clean the cabin, and polish an old-fashioned
brass stove. Breakfast was to be ready by eight o'clock,
and with the assistance of the cook all was prepared.. It
was a great change from home comforts, but my usefulness
in doing odd jobs for my mother came in handy, and it did
not make me feel so awkward as many boys who go to sea.
It began to blow a gale of wind, and it made me feel very
ill. Sympathy there was none. I had to rough it out, and
it was not long before I became thoroughly well seasoned.
In a few days we got into Vidlon Voe, on the N.E. coast of
the Shetland Islands, and brought up in a safe and good
harbour. The captain went overland to Lerwick, to ship
the remainder of the crew, which was to form our complement of forty-five men. We carried seven boats for the
whaling trade. We stayed at Vidlon Voe about a week,
and took in the other portion of our crew. The wind
became fair and we got under weigh and sailed for the
Straits. The anchors were stowed, boats' crews chosen,
and the watch set, and by night we were in the open
Atlantic.

For three weeks nothing eventful came under my
observation. At the end of that time we sighted the
wonderful icefields of the Arctic Zone, and met with many
icebergs. Words fail me to describe the resplendent and
magnificent views which were opened out to me for the
first time in my life, and I did not neglect any opportunity
of coming on deck to look at many most beautiful but
dangerous icebergs. All hands were now ordered to prepare
the boats for catching whales, or anything else which would
produce oil. The day was very beautiful, and the

atmosphere delightfully clear. By noon the lines were coiled in the boats, and the crew went to dinner, one officer and a few men being left on deck to work the ship and look out. I came out of the cabin during the time the captain was eating his meal, to have another look at the great stretch of ice. Something like smoke was rising from amongst the broken masses, and I said to the officer, "Look, Philip, there's some smoke behind that piece of ice." Apparently amused, he replied, "Nonsense, my little man; there are no houses here." Seeing the same thing once more, I said, "Look, Philip, there it is again." Following my gaze he instantly became excited, and shouted, "A fish close to." I was confounded and astonished at the confusion so soon made. All hands rushed upon deck, and the captain went to the mast-head. The call to action had been so sudden and unexpected that the boats had not got the guns in, only the hand harpoons bent to the lines. Without a moment's waste of time the boats were lowered, and in about a quarter of an hour they were away from the ship. The captain called from the crow's nest, "A fall! a fall!" signifying that one of the harpooners had struck the whale. It was my friend Philip who had been so fortunate. We reached close to him with the ship, and ascertained that there were two and a half lines out. The rope was running swiftly round the bollard of the boat and smoking, and the crew were throwing water upon it to keep it from burning. The boats hoist a flag when they fasten to a whale, to show which is the fast boat, and the ship does also. Now the boats were all placed in the direction where the whale is expected to rise. In about twenty minutes she came to the surface, and two more boats got fast, leaving three to lance. It was most interesting to see our boats so dexterously managed. The whale began to blow blood, and in a few minutes the sea, the boats and the men in them, were crimsoned with the life fluid of the great Leviathan. In two hours it was dead, and then came the trouble to get alongside to flense. This took some time to prepare, as we had not expected to see a fish so soon. When all was ready, the crew had half an hour for their meals, after which we proceeded to the work of flensing. The captain rigged me in a blue flannel shirt, tied round the waist with a piece of spun yarn, and also gave me a pair of his sea boots to put on. I was ordered to go into the Malemauk boat. This is the name of two boats laid alongside the whale, and used for the harpooners to place their weapons in during the process of flensing. The Captain jocosely observed to me, "Now, you must always be a malemauk boy in smooth water, as I want to make you a clever harpooner." Notwithstanding, it was too soon to place me there on account of my strength. I found out the benefit of it, however, and, before the voyage was over, I knew all the ways and cuts of flensing a fish. Some may go to sea a life time, and remain utterly ignorant of this very particular operation.

Our first whale was soon followed by another, and we went further north, towards the island of Disco. At this period of the year the sun does not set in these latitudes, and it is light the night through. We arrived safely at the entrance of the harbour of Goodhavn, the principal settlement of the district. The natives came on board with Governor Major Olricks' compliments to the captain. For the Governor we carried despatches from the Danish Government, and also luxuries from the Danish Consul at Hull, and other friends.

The natives upon the East Coast of Davis's Straits, from Cape Farewell to Upernavik, are civilised, and under the Danish Government. They bring for barter small canoes, tobacco pouches, and other articles, for which the crews give clothes, soap, and such things as they can spare in exchange. I took advantage of the opportunity, and got a small canoe and some tobacco pouches for friends at home. We stayed at Goodhavn about one hour, and cruised in the neighbourhood until next day. We fastened to another whale, and whilst in the act of killing her, we clued up the topgallant sails, on purpose to make fast to an iceberg, so that we might be able to flense comfortably. I was on the main topgallant yard, helping to stow the sail, when the captain sang out to us to hold fast. Casting my eyes below I saw the whale under water, coming towards the ship's broadside. She was beautifully distinct, and slow in motion. She slightly touched the vessel, yet the concussion made her tremble as though we had struck a heavy piece of ice. We soon got the whale killed and flensed, and the next day fell in with nearly all the ships from Dundee, Kirkcaldy, Aberdeen, and other ports. On the east side we had killed five whales, in all about seventy-five tons. We proceeded in company with the fleet towards that much dreaded Melville Bay, and reached there after a deal of labour in towing and tracking along the fields of ice, and amongst islands and rocks. When we reached the Duck Islands, and entered the Bay, our troubles began in earnest. We had to track and tow, and saw docks, and we succeeded after four weeks' trouble, but three vessels belonging to the fleet were lost in approaching to the north water, off Cape Dudley Digges. The last neck of ice closed upon us, and there we lay beset two days. The Esquimaux came off to us in their sledges. They were dressed in bear skins, and were very wild. The wind changing, we were liberated, and got into clear water, but a dense fog came on, and, knowing we were not far from land, all precautions were used. The wind was in such a direction we dare not dodge, but were forced to carry canvas, and risk all for fear of being jammed between the ice and the land. In the act of staying the ship we struck upon a high pinnacle island, called Conical Island, or Dalrymple Rock. All hands were called, the boats were lowered, and a warp run to an iceberg close by. We began to heave, and the ship came off amid the cheers of the crew, who desired a good voyage. We made for the west side, and fetched across to Princess Charlotte's Monument, off Jones's Sound. Proceeding south, we fell in with the whole of the fleet, with the exception of the Lady Jane, Superior, and Prince of Wales, which were lost in the Bay. Off Pond's Bay we saw a large number of whales, but they were of a small description. Some ships caught many, but our captain gave orders to his men not to strike a whale unless it was large. These instructions are very proper when you can pick and choose, but now all must be taken which comes first. In this neighbourhood we caught several bears. One was so bold as to come to the ship's side and lick off the blood and oil, during the time the boats were away killing a whale.

Whilst cruising about Pond's Bay and Coutt's Inlet, we caught six whales, which brought our total number killed to eleven large ones. The captain held a consultation with Captain Penny, of the Advice, of Dundee, on the advisability of going to Lancaster Sound, to search for traces of Sir J. Franklin. Both our ship and the Advice had received orders to proceed on this mission before leaving home. We went to Lancaster Sound, and there we found the ice fast from Croker Bay on the north side, to Navy Board Inlet on the south side. We landed coals and despatches upon Cape Hay, and erected a pole. Immediately we got safely on board again, we encountered a very heavy gale. Both ships required the most skilful handling to keep them off the pack ice. The sea was very heavy, and it would have proved fatal to us if we had not been able to weather the pack. We went through this awful storm splendidly, and it reflected the highest credit upon our respective commanders and their crews. The gale

B

moderating, we sailed south and joined the fleet.   They
had not killed any more whales from the time of our
leaving to rejoining them, on account of the gale of
wind breaking the land floe up.   We got further south, and
captured two more large whales off the North Cheek of
Scott's Inlet.   These made thirteen fish altogether, of 160
tons, and completed the quantity we required.   With light
hearts, and joyous anticipations of home welcome, we filled
the spare water casks and secured the boats for the return
voyage.   Being the first ship bound for home, we received
letters from the crews of the other vessels for their friends,
for although all had been fortunate in killing fish, they
intended staying a little longer to fill up if possible.   The
last to leave us was Captain Dring, of the St. Andrew,
belonging to Aberdeen.   His vessel was nearly full, and
was going to follow us in a few days.   Within forty-eight
hours afterwards, however, Captain Dring met his death in
a shocking manner, being accidentally shot.   The
melancholy news was received, shortly after his letters had
been delivered, by the lady to whom he had intended
being married when he returned home.   So sad an accident
was a great shock to all who knew him, for he was a fine
young man, and greatly beloved.

   We experienced a favourable passage across the Atlantic,
but encountered adverse winds when nearing the Shetland
Isles.   We therefore bore up for Long Hope, in the
Orkneys, and there came to anchor and discharged the
Shetland men, who went to their homes in small fishing
boats.   We received, at Long Hope, the distressing tidings
that cholera had visited Hull, and committed frightful
ravages whilst we had been away.   We were told that
nearly the whole of the inhabitants had been swept away by
this awful scourge, and that pits had been dug in which to
bury the dead bodies.   Our hearts sank within us.   We
had looked forward to a joyful home-coming, and here we

were, still far away, filled with unutterable dread lest any of
our dear relatives and friends should have been stricken
down in our absence. The following day a favourable
breeze sprang up. We had taken in a good supply of
fresh meat and potatoes, and we got under weigh and
sailed homewards, very much depressed, and with gloomy
forebodings. Off Peterhead we landed letters. The
fishing boats were informed of the loss of the "Lady Jane,"
and "Superior." Both ships belonged to Peterhead, and
the report of these casualties made the men very sad, as
most of the people belonging to that place, were, at this
period, deeply interested in the whaling and sealing trade.
We only allowed ourselves one hour's stay, and we resumed
our journey homewards. The nearer we approached old
Humber, the more anxious we became respecting the welfare
of those near and dear to us. Wind and weather favoured
us, and we arrived in the river a few days before Hull Fair.
Our ship's reappearance in the roads after so many months'
absence, was quickly reported in the town, and crowds
arrived on the piers to see us land. There was general
handshaking, and on our part eager enquiries after our dear
ones. Not one of us had lost a relative, and the stout
hearts on our ship, who had cheerfully braved the regions
of the northern climes, were thankful to Almighty God for
His mercy, in preserving from the plague all who were near
and dear to them during their absence. I was the last to
leave the ship, and I shall never forget the meeting with my
dear mother after seven and a half months' separation.
Both had gone through great danger during that eventful
period, and her joy and affection at having me by her side
once more will remain in my memory till my dying day.
My sister's heart, too, was full, and we rejoiced together at
our re-union after so many months' separation.

The discharging of the blubber and whalebone followed
our arrival in port. It was in a terribly dirty state, but

everybody was happy at having made so lucrative a voyage, and no one cared for the dirt. The casks were hove out of the hold, put into lighters, and conveyed to the Greenland yards, which to-day is a thing of the past. There the blubber was boiled, and the whalebone scraped and cleaned ready for market. The amount realised by the sale of the oil was very large. The ship was laid up for about six weeks. During that time she was cleaned, and we received our prize money for the attempt to succour Sir John Franklin. About £1000 was divided between the "Advice" and the "Truelove" for their services, and my share was £1. During the winter I had to be on board from daylight to dark, and keep the ship clean. The work was very heavy, for I was the only apprentice on board. The winter passed, and nothing of any note occurred respecting myself. On January 20th, 1850, another mate came on board, and was followed by some of our old harpooners, to refit the ship. Our new mate was quite the reverse of the one we had last voyage. He treated me kindly from the first. The other never omitted to annoy me in any way he could. I had, however, many very good friends on board the ship, and I therefore submitted to his ill-mannered and unjustifiable conduct.

CHAPTER II.

WE mustered the crew about the 6th March, 1850, to try our fortunes once more. I was now promoted to the forecastle, and a steward took my place in the cabin. We sailed from Hull amid the hearty good-wishes of all, and arrived safely at Lerwick. The harbour here is most commodious, and a fleet of men-of-war could anchor in it in safety. Natives came from the mainland and the island of Brassy, with eggs, fish, potatoes, and other commodities, supplies of which we purchased to add to our stores of food. We shipped the remainder of our crew, and after lying there four days awaiting a favourable breeze, we got under weigh with a scant wind. About 300 miles from the land we encountered strong winds, in consequence of which we were seriously delayed in getting to the fishing grounds in good time. The whale trade was perhaps the most speculative of any at that period. Mates were only paid £2 15s. per month. This was supplemented by seven guineas bounty, 8s. per ton for oil, 30s. per ton for whalebone, 21s. per fish when the longest blade was six feet long, and 10s. for the first harpoon. Harpooners received pay at the rate of 30s. per month, seven guineas bounty, 20s. per ton for whalebone, 8s. per ton for oil, and 10s. striking or first harpoon. Boatsteerers obtained £2 10s. per month, 2s. 6d. oil money, and 5s. per ton whalebone. In the case of a ship coming home without any whales or seals, harpooners would be in great straits, as their 30s. per month was drawn by their wives for the maintenance of themselves and their families. The men had also to find their own tea, coffee, sugar, tobacco, and other articles. In the present day whalers have better wages, though less oil money. Steam has superseded sailing vessels, and the voyages are less hazardous and more lucrative.

To return to my narrative. Though we were badly delayed by strong winds we did not despair. At the first fishing grounds we did not see so many fish as in the last voyage. The state of the ice made it more difficult to thread our way, but we struck two whales, and after being fast some seven or eight hours, the harpoons drew. Unfortunately we lost both fish, the ice being so closely packed that the boats could not get to them. The season on the east side was drawing to a close, and we began to look for a north passage through Melville Bay. We got no further than the Vrow Islands and Upernavik, the northernmost settlement belonging to the Danes. We failed to get an opening northward, on account of the lateness of the season, and a consultation took place between the captains of the ships who were sailing in company. It was resolved to try the south end of the ice, and get a passage north on the west side. We tried all the bights down southward until we came off Cape Walsingham, on the west side. Getting through the pack there, we succeeded in reaching the so-called west water, and with a light fair wind we ran up along the coast, distant about ten miles. We were stopped by the ice several times for days. Ultimately we arrived off Home Bay, but it was now September, and the nights were increasing in darkness.

A run of whales came down, and during the evening we harpooned three large ones, and very fortunately succeeded in killing them. I well remember the night. It was still and calm, the moon was at its full, moving in stately majesty in unclouded sky Its soft beaming light fell upon ice and sea, producing wondrous phantasies, and inspiring admiration and awe. Never before had I seen such marvels of nature in the Arctic Circle. The scene was grandly sublime, and impressed all not actively engaged in securing the whales. At daylight we had all three killed and alongside the ship. We began to flense, but the operation took a long time, the men being weary with their night's work. At such times men were not allowed to give up until they fairly dropped. A fresh northerly wind springing up, we ran for a place called Cape Hooper. Here we made off our blubber, or plainly speaking, cut it up and stowed it in the casks, which took three days. We sent our boats towards the coasts to look out for whales coming south along the land. This is called by the men " Rock Noseing," and is about six miles pull from the ship, and a cold and hazardous duty. All hands were generally called before daylight. They got their coffee and manned their boats for the day. If not successful they had to return to the ship in the evening, and hope for better success another day. By the time the crew turned in, they were very tired with their dreary search throughout the day. This business was generally done in the latter part of the year, when the nights were dark, and the weather treacherous, and unsafe for ships to stay out among the icebergs and broken masses of ice. The harbour we were in was one of many anchorages at the back of islands, or small bays sheltered from the sea. Such places are numerous, and well-known to those visiting them, but a stranger could not find his way to them, as there are no charts or any other guide but previous experience. Each ship generally sent six boats away, with provisions to last a couple of days, leaving one or two boats on board, in case of a fish rising near to, as sometimes turned out to be the case. The remainder of the crew were employed in watering, bending sails, and other necessary jobs for the passage home.

In this harbour there were seven ships, each of which had obtained several whales. Other vessels did not come in, and were waiting outside in the hope that whales would make their appearance at the headlands, which they were accustomed to do. One night it blew a strong wind from the N.E., and the ice came from the north, rapidly filling the harbour, creeks, and small bays until the whole assumed the appearance of a solid field of ice. We were unable to lower the boats and get away, and several captains travelled to the top of a high mountain, to get a view of the state of the ice to the northward. After seven hours' absence, they arrived back with the news that nothing but ice was to be seen to the northward and eastward. The ships outside were about ten miles to the southward, and under weigh, but we seven were here like rats in a trap, and it was a sorry look-out for us. The masters held a consultation, and it was proposed that about twenty men belonging to each ship should stay by their own vessels, and the remainder make an attempt to reach the ships in the offing. Next morning all hands were called, and after breakfast they proceeded to make preparations to leave. The men who were to remain by the ship gave their messmates a helping hand in taking the boats over the ice. They reached a low point of land about two miles distant, but found it too difficult to launch the boats with the lines in. These were therefore coiled on the land, and the light boats launched a little further on. The parting between the two bodies of men was anything but cheerful under such circumstances. Each encouraged the other with a true British cheer, which inspired fresh hope for the future. It was dark when our men returned on board. They were thoroughly worn out with hard toil. During the night it began to blow and snow heavily, and we became extremely anxious for the welfare of those who had left us. With the light of another day, we went to drag the lines over the ice to the ship. Night again came on with our work unfinished, for we had only got one boat's lines out of four, and we were very tired and wet when we turned in.

A startling change occurred in the night. The wind blew a heavy gale off the land, and the ice drifted out of the Bay. Our anchor had got foul of a rock. We could not heave it up, so we slipped it. The night was very dark, the wind blew heavy, with squalls and showers of blinding snow. The channel was narrow and shoaled on the north side. On the south side it was rocky, which made it very difficult and dangerous to navigate. These serious dangers were greatly increased by the close proximity of the ships to one another. It was indeed a most unfortunate predicament to be in. By the exercise of great care we got clear of the land a few miles, and lay to until daylight. We were anxious for the lives of those who had left us, for we knew that if they had failed to reach the other ships in the open, their safety would be greatly imperilled by exposure to the dangers which had set us free. Daylight broke at length, for which we were thankful. The hardships we were enduring tried the strongest constitution. Cheered by the light of another day, and more moderate weather, we ran to the southward, but saw no signs of boats or ships. At Cape Searle, however, our hearts were gladdened at sighting our consorts not many miles from us. We spoke the first vessel as soon as practicable, and we were told that all the men were safe, but had lost both clothes and boats, having had to leave them upon the ice and run for their lives before the gale and darkness set in.

The reader may perhaps wonder why the men were obliged to leave us. The ships were all well provisioned, but not sufficiently for wintering in Davis's Straits. When the seven ships were pinned by the ice in the harbour, there was not sufficient food on board for the support of forty-five men, which composed each crew, for eighteen

months. The supply was adequate enough for twenty men,
and therefore the remainder had to depart for fear our fleet
was unable to get into the open again. We stayed off
Cape Searle until noon of the following day, when we bore
up with a fair wind for home in company with all the other
ships. During the time we had been out, our patience and
fortitude had been very much tried by the blighting of our
hopes of a prosperous voyage. When we got our three
whales we had a fine prospect before us. A short time
sufficed to reverse the picture. After a very stormy passage
we arrived at Lerwick, and there discharged the Shetland
men. We continued our homeward voyage, and were
richly consoled for our very unlucky voyage by the warmth
and cordiality of our welcome at Hull, which we reached
the last week in October.

CHAPTER III.

ON the 12th March, 1851, we were once more fitted
out for the Arctic seas, and left Hull under very
favourable circumstances. At Stromness, the principal
harbour in the Orkney Islands, we took in our full
complement of men. The seamen belonging to these
islands are similar to the Shetland men. At the period
of which I write, a great many had been in the Hudson
Bay Company's service, and were adapted for the work
through the many privations they had experienced in the
Company's employ. They were a nice, quiet, and
temperate people, and amenable to good discipline.
Those who had been in Hudson's Bay, particularly, were
older and better men. We always had to take many young
hands, but all were good boatmen as well as farmers, a
patient hardy race of men. Articles were signed and all
came on board. There was not a single case of drunken-
ness among them, nor have I ever seen one, either from
Shetland or the Orkneys. We got under weigh, proceeded
through Hoy Sound, and were soon in the Western Ocean.
We passed the outer islands with a fine fair wind which
carried us well across the ocean towards Cape Farewell.

The first ice was made by us in what is now called
the old S.W. Lat. about 62½ N. Long. 54 W. We
went along the ice edge until we arrived off Holsteinberg
or Weideford, as it is commonly called. There the ice led us

inside several rocks and islands, and we grappled our way
close along the land until we came to Riff Koll, when it
began to blow and snow. The ice set on to the land, and we
had a very close shave between two high rocks, which gave
us barely room for the ship to pass. We got past the rocks,
and the place then opened into a wider space. We made
fast to the land floe a few miles from a settlement. The
natives came off in their sledges to barter, and the Governor
visited the captain, with whom he had been long acquainted.
We were glad that we had got into such snug quarters,
though the ice was piled over the low rocks on the outside, and we were nicely landlocked. The natives brought a
violin, and the 'tween decks were cleared away for dancing.
The females are great dancers, and are not fatigued after
ten or twelve hours' enjoyment of such exercise. The
few hours they remained on board passed very merrily.
Next morning the wind and weather changed. We cast off
from the land floe, and bade adieu to our friends. One
boat was sent ahead to look out for rocks, and the others
assisted the ship by towing, the wind being light.

Getting outside we had more wind, but found ourselves
among some loose floes to the north and west. A whale
was sighted from the crow's nest. All hands were called
and chase given. One of the boats got fast to her, and
in four hours she was dead and alongside our ship.
This was our first whale. We had no ship in company, and
we thought we were going to be very fortunate, as we do
not care for too many to be about on fishing grounds.
The following day we gave chase to several whales, but a
lot of young ice surrounding us we did not succeed in
getting fast to any. One was fired at but missed, which
very much displeased the captain. Five more vessels made
their appearance to the southward, and the ice opening
towards the northward, we made sail for the island of Disco.
This is a noted place for whales when the ice lies favour-

able in an early season. We were no sooner close to the
land, called by the whalers Black Land, from its appearance,
than a great many fish were seen. We struck one and killed
her, then made fast to an iceberg to flense comfortably, and also to keep us from drifting, as a sailing vessel is
nearly unmanageable with a dead whale alongside. We
finished our work and gave chase to several more which
came near us, but did not succeed in striking another
that day.

The weather was calm and clear next morning. The
whales close in shore were very numerous, but we could not
get near them. We pulled, sculled, and used the paddles
to no purpose, for they are very quick of hearing, and being
within a mile from the shore it was supposed the high land
attracted the sound under water, and alarmed them.
Our oars were muffled, and the boats pulled without any
noise, but our efforts were not rewarded by a catch, and that
day passed without anything eventful. A breeze sprang up
in the morning, and a whale rising close to the ship, we
lowered two boats. As she rose again we struck her.
The harpoon was no sooner in than she swept the boat of one
side of its oars and filled it with water. Nothing daunted,
we got another harpoon in, and soon killed her with lances.

Sometimes we put three, or perhaps four, harpoons in
before we begin lancing, according to circumstances. We
now killed our third whale, and continued to ply
about, first in the offing and then inshore, but no fish were
to be seen. We sailed north towards Hare Island. There
we lay at an iceberg a week, close to the land, and watered
from a large stream which ran down the valley. This island
projects out, and stops the ice from drifting south,
especially when many bergs ground a short distance off.
It became calm at last, and the ice opening, we began to
tow in company with twelve other ships. For eight hours
we were thus employed, and then reached open water.

A gentle breeze rose, and we went off to the northward towards Black Hook, on the north side of N.E. Bay, commonly called Jacob's Bight. We cruised about for a day or two, but did not meet with anything. At length a whale was sighted close to, and all hands called to action. She was struck or shot with the gun harpoon on making her second appearance. The harpooner of boat No. 2 delivered his harpoon, and in an instant the huge animal struck the boat, capsizing it and throwing the crew into the water. She lashed her tail and fins amongst the poor fellows, dealing fearful blows here and there. The men overboard clung to oars, or anything they could get hold of, and another boat gallantly picked them up. One poor fellow was quite dead, although he had hold of an oar and was floating. It was believed he had been struck by the animal and killed instantaneously. All of them were conveyed on board, and three more harpoons were put into the whale, and she was speedily killed by lances.

The death of our poor shipmate was felt by all who knew him. He had fallen a victim to a casualty which threatened us all, and in accepting the service we were bound to take the risk. After the whale was flensed we cruised about for two days, at times having all hands away, and sometimes two boats on the watch. The season was on the wane, and a fair wind springing up we ran to the northward. At the Vrow Islands, close to Upernavik, we came to a heavy block of ice. The out-lying islands and rocks kept the ice fast, and made a passage through impossible without the aid of an easterly wind or calm. We sought a passage inside, but all was blocked, and we therefore made fast to an iceberg close to an island. In the meantime our carpenter had made a coffin for the unfortunate man who was killed. We had been waiting the first opportunity of burying him on the land. All hands were called, and the crews from the other ships invited to be present at the funeral. The scanty soil was only a few inches deep, and his grave was speedily dug with crow bars, for shovels were of no use. The funeral procession was most solemn and impressive. All the ships had their flags half-mast, and about thirty boats, each containing six men, towed the one in which was the coffin and its occupant slowly towards the land. The doctor read the funeral service, and we covered the wooden shell with large stones, placed in such a position that they did not rest upon it. A wooden head board with the name, age, and birthplace of the deceased, and the ship to which he belonged, marked his burial place. I saw it several years after, and it was in good preservation. I may add that I have seen other head boards, fifty years after their erection, in good condition, although bleached quite white by the weather. Many poor fellows are now resting in this inhospitable and sterile country. No gentle hand is there to place a flower upon their lowly graves. They died in procuring the necessaries of life for their wives and families, and in this they were doing their duty. Their sacrifices were great, and such brave hearts will not perhaps be forgotten in the great roll-call by the recording angel. No grave that I have ever seen there has been disturbed by either bears, foxes, or wolves. Can such a thing be said in our own country, where the flowers on the graves of our dear ones are constantly being plucked by sacrilegious hands.

Two days after, came a calm, and the ice began to open in all directions. Boats were speedily manned in order to tow the ships through the intricate passages of the islands to the northward. It is a dreary and slow process, but when a few ships are together it becomes lively and amusing. One ship tows against another, and the crews strike up some song or chorus, which has a very fine effect amongst the numerous islands and icebergs, the echoes coming from every direction. We continued towing, tracking, and taking advantage of the light airs which prevail at this season for short periods. These, however, cannot be depended upon, for sometimes a sudden storm will rise from the S.W., with rain or thick snow, which in a few hours reduces us to despondency. We arrived off the Devil's Thumb, a little north of the Duck Islands, and forming the south side of Melville Bay. The Thumb is a very high pinnacle or headland, in the shape of a thumb, and such headlands in this region have singular formations. One is called Kettle Bottom-up. It appears to resemble a pitch kettle in one direction, but from another view it is called Sugar Loaf Hill. Such names are given by the crews who visit this part of the world. Charts are of no avail here, and all must derive their knowledge of sunken rocks and other objects by experience.

We waited a few more days for the loose floes to open, and towed and tracked several miles. The ice began to close, and a breeze sprang up from the S.W., which increased to a gale. The ships had secured the safest places for the time, and we had sawed docks in the land ice, which was a long and tedious job. Some had got into their docks, but others had not, when the loose floes swept on us and completely buried the land floe. Dismayed by the crash, all hands launched their boats a distance from the vessels, and filled them with provisions in preparation for the worst. It was blowing and snowing heavily. One of the ships was cut clean through by the ice, and her masts laid on it. The awful squeezing and crushing lasted about six hours, when the ice became stationary. Plundering of the unfortunate ship which had been wrecked, followed, and I am sorry to state it led to a most disgraceful exhibition. Some of the men tumbled the rum puncheon out of the cabin. The head was knocked out of the cask, and tins and boots were dipped into the rum, which was quaffed until a number of them were intoxicated, and strayed away in different directions. The gale moderated, although it was snowing and raining. Parties were sent from the ships to look for those who were absent, and providentially all were found, as several bears had been prowling about. Some of them were discovered asleep behind hummocks of ice. We were not a moment too soon in the work of rescue. I heard afterwards that several were cripples for life. Some were attacked with rheumatism, and no doubt shortened their lives by years. It is sad to think how men will risk their lives for the sake of beastly indulgence at a wreck, as I have seen some do. The following morning the ice slackening, somebody set the wreck on fire, and she soon disappeared.

During the evening it became calm, and the ice opening, we got the provisions back on board, and began to track the ship along the land floe. Coming to a block, we cut docks. Generally two ships are placed in one dock, because it is sooner done. Sawing a dock is a difficult task, and great care must be taken to get the lines straight, otherwise there is a probability of having nearly as much to cut over again. It must be wider at the mouth than at the end, and sometimes the block will come out in one piece. If it is not sawn with care another cut must be made called a jib piece. Ice saws are from twelve to sixteen feet long, worked with triangles and bell-ropes, and manned by about sixteen men. The work is usually enlivened with songs to cheer the gang. I have seen blasts of powder put into the middle piece of ice, which saves much time and labour. We had no sooner got the ships into their docks, than the sky began to make up in the S.W., a sure forerunner of a gale from that quarter. All preparations were made. The boats and provisions were again on the ice, and in four hours from getting the ships in comparative safety, the gale came upon us accompanied by blinding snow and rain. It was more

C

severe than the previous one, and although all hands had been up so many hours and heavily worked, they were very sprightly, the excitement having stimulated them. This gale lasted twelve hours, and we were glad the ice had not much room to give it force in the first place, or else few of the ships would have stood the pressure of ice upon them, which was very heavy, and made us quake. For several days we were firmly jammed. The ice was ploughed up many feet around us, and the outlook was not very promising. At last a general calm came, and we were liberated once more, and started again. We made but little progress, and several times had to make fast for a few hours. Eventually we got as far as the Sabine Islands, about thirty miles distant.

The season advancing, and there being no prospect of a passage through the Bay, a consultation was held by the captains. It was determined to retrace our way back and seek a south passage. Being the last of July, there was no sign of water to the northward. We ought to have been in Pond's Bay, on the west side, before the middle of June or thereabouts. All hands were called to tow to the southward, the ice opening in that direction, and in less than 15 hours we were in the south water. With a breeze of wind we could work the ships. It had taken us five weeks to get where we had been. We all pursued different directions, some trying every bight, and others making progress to the southward, until we arrived off Cape Dyer in 66½° north latitude. The ice being loose and broken up, we worked our way in a dense fog. It cleared up, and we found we were not far from the west land, off Dyer's Cape. We ran to the northward, with a nice breeze, in clear water, the ice being about ten miles distant from the land in some parts, and more in others. We arrived off Cape Hooper, came to a block, and knocked about among the loose ice in the hope of seeing whales, although the time was too soon for them so far south. A fine breeze came off the land and blew the loose ice from the land floes. We got as far as Cape Kater and sighted a whale. It cheered us up for a moment, but we did not succeed in getting near to her. Every day our boats were away, sometimes for hours, but without success. We shot a few bears during the time we were cruising about, and a walrus.

Some people may think that when whales are not found there is no work to do, but I can assure them there is less work in a full ship than with a clean one. When not on fishing grounds, the crews are constantly employed under the boatswain, wind and weather permitting. On board things may not be so pleasant as they ought, for there are people who will show their ill-temper when they can. Boats manned have been sent away for hours when nothing has been seen, especially when it was a cold, bitter day, on account of the captain having heard some one grumbling or saying we ought to do this or that. If the captain was in the crow's nest, every word could be heard that was spoken below, especially when the wind was light. Discipline was very strict on board in those days. Our articles were signed by us to kill, slay, and destroy any living animal or animals in the waters by day or by night, at the order of the captain.

September now being in, we made the best of our way towards Cape Searle or Malemauk Head. This is an island with a very perpendicular cliff. At the back part is good anchorage for a few ships. It has only one way in, and that is from the north. The south entrance has a bar across with only eight to ten feet of water over it, otherwise it would be a very safe place. When the ice comes down from the north, great care has to be used, and a good look out kept, for it sweeps round Merchants' Bay into this place. During the latter part of the season, when the north winds prevail, and the ice once gets into the harbour, there is very little chance of its coming out again that year.   If a ship is not careful to get close to the land, the anchor might be dropped in 120 fathoms, for it deepens very suddenly.   The island projecting well out, whales frequently make their appearance, and it used to be a favourite place for ships.   Many good fish have been caught here.   When the boats left the ships some went round one way and some the other, and met on the north side.

We stayed here about a week, and the ice being seen coming rapidly south, we got under weigh, and went a little further to the southward, to a place called Durban, another fine harbour when free of ice, but a long way for the boats to pull to the outside.   This harbour has also very peculiar characteristics.   Ships have been driven out by the ice coming from inland.   It was a mystery how the ice came that way, as none was seen at the outside.   In 1854, three boats belonging to the Eclipse were suddenly beset by ice on the north side of Merchants' Bay, the crews being forced to leave them, and nine days later the boats were found drifting out of Durban harbour, which solved the problem.   A full account will be seen in the sixth voyage. The whole of the west side is nothing but islands, bays, and fiords, which nobody knows anything about, and which are not worth the trouble to explore.

Not having seen any whales, we came down to Dyer's Cape, but no harbour being there, we sent our boats inshore.   The ship was under weigh, cruising off and on, and the full watch having to be set during the whole night, our work was very laborious.   The men through not getting rest, were liable to go to sleep in the boats.   Fortunately, we had only three days of it.   We ran south towards Exeter Sound.   This place has an inner and an outer harbour.   The inner harbour is seldom used on account of the long pull out to sea.   The outer one is exposed

to the east wind when it blows from that quarter, which
sends a heavy swell in when there is no ice outside
to shelter it, hence the bad reputation of the place.
Our captain determined to go to Cumberland Sound.
The ice does not come into this Sound with northerly
winds, but drifts about forty miles from the mouth of it.
We anchored at Niatlik, and sent the boats away as usual.
Off this place is to be seen Black Lead Island, so called on
account of the quantity of that metal found upon it. I
have found pieces so fine that one would think it had come
out of the manufactory.

The weather at this time of the year is very treacherous,
and more particularly about this place, on account of the
high mountains and deeps fiords. We had to keep a good
look out, and the plan generally adopted was to send the
boats away at daybreak, with instructions to return early in
the evening. One day the weather was lovely, and three
boats pulled to Black Lead Island, at the outer edge
of the reef. About three in the afternoon, we got fast to a
large fish with the gun harpoon. She proved a very
dangerous customer through being in shoal water. We
put another harpoon into her quickly, and the other boat
sent two or three lances into the vital parts. After a final
flourish, which is generally called the dying flurry, she
succumbed to us.

The other boats, having seen our jacks flying in the
offing, came to our assistance. It is a long and tedious
process, but as night was approaching, everybody pulled
with a will. The captain had sighted us from the island,
and sent the seventh boat to our assistance. Lights were
also kept burning on the island to guide us in. About
two in the morning we got alongside, having had nine
miles to tow. We were all ready for a rest, and some of us
were very wet the whole time we had been employed in
killing the whale. After securing it to the ship, we had a

good four hours' sleep.   Flensing followed, and the carcase
was towed to shore at high water.   It was dry at low water,
and made a great feast for the natives and their dogs.   The
skin of the whale is very good and nutritious.   It has a
taste of mushroom after being cleaned and steeped in salt
and water for twenty-four hours.   When boiled about
twenty minutes, and cut into small pieces, and eaten with
vinegar it is a good antiscorbutic.   The natives eat it raw, but
it was not so palatable to us.   Therefore we drew the line
at that.   I have eaten it with a relish, and on my last voyage
to this country, when mate of the Diana, of Hull, my boat's
crew used to take about eight pounds every day during the
last two weeks.

The following morning we sent four boats away, and the remainder of the crew put the blubber in casks.   This made our
fifth whale, from the whole of which we got about seventy-
five tons of oil, and four tons of whalebone.   Stormy
weather came on, and it was freezing keenly.   Notwithstanding, we expected getting another whale before we left.
One evening a boat belonging to the Regalia, of Kirkcaldy,
struck or shot a fish.   It was coming on dusk at the time, and
the boats were far out at sea.   When darkness came on,
the wind increased, with blinding snow, which added to
the misery of the poor fellows in the lonely boats.   Tar
barrels were burnt upon the island, but they were not much
use, on account of the obscuring snow, and it was a most
anxious time for those on board the Regalia.   Next morning
was clearer and more moderate.   We sent our boats away in
search of their's.   They were met about five miles from the
harbour in a most deplorable condition.   All were quite
worn out, wet to the skin, frost-bitten, and hungry.   We
gave them provisions and some rum, and took part of
them in tow.   From their statement it seems that when night
came on, they had only got two harpoons in the whale, and
the sea was increasing, but was very phosphorescent, so

they could easily see her in the dark. They put two more harpoons into her, and commenced to lance, but the darkness made it very difficult and dangerous. About ten o'clock they got her killed, and commenced as soon as possible to tow. The sea being so heavy, it was a dreary and tedious process. After struggling with her about four hours, they were obliged to let her go, as it was impossible for a boat to live entangled by a dead whale. They then laid to with the boat's head to the sea, waiting for the weather to clear, and for daylight. Thus it was we found them. Some of the men were unable to use an oar. Their bravery merited a better result than fickle fortune gave them. Eight of the crew did not go to sea again, and three died during the following winter at home. Others were crippled for life through losing their toes. Such were the perils of whaling in the days of sailing ships. With powerful steam ships the work is play now to what it used to be. When a boat gets fast to a whale the vessel will have steam at hand, and will perhaps buoy his cable, and in a short time steam close to his boats and tow the whale into the harbour, whilst his crew are getting their coffee, and being otherwise refreshed. After the incident last related, the bay ice began to form fast in the little creeks and bays. It was therefore thought expedient to make sail for home.

The captains of the five ships lying in the harbour had a consultation, and it was decided to leave next day. We supposed we were the last ships in the country. A fine, fair northerly wind blowing, we all got under weigh and proceeded down the Gulf. Whales were seen inshore amongst some bay ice, but it would have been useless to have attempted to go to them. We parted with the other ships at the mouth of the Gulf, and steered more along the land until it became dark, and then set our course for home. After a moderate passage we sailed close to

St. Kilda, Butt of Lewis, and the other islands of the Hebrides, and arrived at Stromness, received our letters, and landed the Orkneymen. The following day we made sail for Hull, and arrived home without any further incident, where we found all well.

CHAPTER IV.

ADVERSE EXPERIENCES—ANOTHER DEATH IN THE ARCTIC
SEAS—SIR JOHN FRANKLIN EXPEDITION—AN AWFUL
SQUEEZE.

ABOUT the middle of March, 1852, we commenced
another voyage to Davis's Straits, many of the old
crew being with us. The preceding winter had been spent
in self-improvement. My spare evenings were passed at
the night school, which was provided by the owners for
their apprentices. I have known about one hundred and
ten apprentices attending the school during the winter
evenings, and it was very beneficial to all. We left port
with the assistance of a tug-boat, towed down the
Humber, and anchored in the Sunk roads for two days,
as a fresh northerly wind was blowing. When it changed
we got under weigh, and went to the northward, duly
arriving at Stromness, in the Orkney islands, where the
winds from the westward detained us a week. We
were in the company of several more ships, and when the
wind became fair, there was a great commotion on the
land and amongst the ships getting their respective crews.
All being on board, topsails mast-headed, and the anchors
hove up, we sailed towards Davis's Straits. It was rather
late when we made the ice, in consequence of adverse
winds. This time the ice led us well south into
the land, off Sukkertoppen, and we made all progress
to the northward. The island of Disco was reached in
safety, and having got our lines coiled, we were ready for anything which came in our way. We cruised about several days.

At last we got fast to a fish, and soon had her despatched. We began to flense, but during the operation the wind, being on the ice, drove us a mile into the loose pack. This could not be avoided, as a ship with a dead whale alongside is not very handy. We were beset two days. During that time we had the blubber put away, and made ready for further catches. It became calm, and the ice slacking, we towed into clear water. All the ships had already gone north, and no more whales being seen, we followed. We reached as far as Hare Island, and made fast to an iceberg, close to the land, when a heavy gale sprang up. We had six warps made fast to the iceberg, yet notwithstanding there was great danger of their parting. Luckily they held on during the storm, which lasted twenty-four hours, all hands being in readiness. We were surrounded by ice, and were afraid of the berg floating, and putting us on the weather side. However it kept aground. We were only in 20 fathoms of water, and steep to the land. A boat's crew went on shore to take a view from the top of the island. They came back with the news that the other ships were a long way north, and all had made fast to icebergs. Very little water was seen. This was very disheartening, as we had expected by this time to be in the Black Hook water, amongst whales. We consoled ourselves with the reflection that a few hours' calm might make a great alteration. The ice slacking inside of the Waigat Straits, and towards the island of Disco, all hands were called to tow. A current drifted us near to the island, and brought the ice upon us. We again made fast to an iceberg close to the land. An hour afterwards a poor young fellow died on board. He had been ill all the voyage from Stromness; he was consumptive. The doctor had been very attentive to him, and he received every assistance from the crew. The carpenter made a coffin, and carved a headboard out, and we called all hands to bury him on the land. The north part of Disco is a low sloping point, and here his grave was made. The funeral service was gone through, and the coffin well covered with large stones. The sad ceremony over, we returned on board, leaving him far from his native shore on a lonely bleak spot. After setting the watch, some of us were about to turn in, when all hands were called to tow. This time we succeeded in getting to Four Island Point, on the south side of N.E. Bay. The governor and natives came off with barter. This is a very small settlement, and was at this period under the care of a Danish governor, married to an Esquimaux. The law was strict upon these marriages. If a Dane wished to return to his native country, he must take his wife with him, and be worth a certain sum of money. Under such conditions, not one in twenty could or did return, but had to remain in the country.

We stayed here two or three days until we got liberated, then we stood off to the ice edge, to the westward, and saw several whales. We got fast to one and had her killed in five hours. She got amongst a lot of loose ice ; it was difficult to get our boats near her, but when she came to the outside we finished her. After cruising about several days, and not seeing anything, we began to use our usual means to get round the north end of the ice, through Melville Bay. With much trouble we reached the Duck Islands, at the entrance of the bay, and there waited a favourable opportunity for the ice floes opening. Calm weather is the best time for it slacking, but breezes set it in commotion, except a light N.E. wind, which sends the loose floes rapidly from the fast or land ice. It is very astonishing what effect a light air of wind has upon such fields of ice. Our fleet consisted of about twelve sail. One evening the ice began to open. Everybody was up and on to the ice with their tracking belts, cheerily pulling their respective ships along. We had to follow in a line whilst tracking along the floe edge, and had several slight squeezes on this occasion, but no signs of a breeze from the S.W.

Our fleet was now joined by the large exploring expedition for Sir J. Franklin, under the command of Sir Ed. Belcher and Capt. Kellett, two old tars who had seen a great deal of service, especially the former, in the China seas. He was the dread of the China pirates, and swept the seas of them. We succeeded, by tracking and towing, in getting sight of Cape York, north part of the Bay. Heavier fields of ice were met with. A very sudden and heavy gale sprang up from the S.W., with snow and rain. All the ships were close together. There was no time to saw a dock, but each had to lie and take their chance. We launched our boats, and put into them provisions and clothes, and left the ships, prepared for what might follow. The ice came crushing and tumbling upon us, and all expected the vessels would be wrecked, for it appeared as if nothing short of a miracle could save us. The American ship, the McLellan, of New London, fell over on her broadside, and lifted out of the water, about nine feet : the ice had cut through her. The scene was intensely exciting, and our minds were full of gloomy apprehensions. With the exception of the captains, everybody had left the ships. Each of the masters was standing on an elevated part of his vessel giving instructions through speaking trumpets, and firmly resolved to be the last to leave. The ships began to crack and groan, and give sudden jumps. This was hailed as a good sign, for it proved that the ice had got under them, and that they would have a better chance of surviving the pressure. At intervals there would be a lull in the movements of the ice. Another sudden squeeze would follow, making the crews anxious for fear their own ships would be the next to be wrecked. This dismal state of things continued for eighteen hours. The crew belonging to the expedition, during the time, gave great assistance by putting blasts of powder in the most critical cases, and relieving the greatest pressures. There is no doubt that their timely aid saved some ships from sharing the same fate as the McLellan.

The gale ceased at last, and the weather became again calm. The crew of the wrecked ship distributed themselves amongst the other vessels, the captain and his boat's crew coming on board our ship. Then came the bustle in getting the boats, provisions, and clothes nearer to the ships, and we were thankful we were able sleep on board. There had been no rest during the gale, for an open boat is not a pleasant place to sleep in, especially when it is blowing, snowing, and raining. The ships had a most singular appearance. Some listed one way, and some another, and all were headed in different directions, giving one the impression that they had been pitched out of a bag on to the packed ice. The McLellan became a total wreck. Our own ship had not had the same heavy pressures on as the other ships. In addition, she was much stronger built, and able to resist more pressure. Next followed the petty plundering of the wreck. The captain of the McLellan sold her to Sir E. Belcher for one hundred pounds. When the marines went on board to take charge, it was amusing to see the plunderers run. Some had a bit of canvas or rope with them, not worth sixpence. The wrecked vessel was upright, leaving her cabin bare, and also her 'tween decks. The expedition took everything that was necessary for themselves. One morning Sir E. Belcher went to the wreck. The ship's name was scratched out, and somebody had painted in its place, "John Bull's prize, by Ned Belcher, the pirate." I was close to Sir Edward, delivering a message to him from our captain, when he read the words. His annoyance at the insult was such that if he could have found the culprit he would have triced him up, and given him a round dozen.

The season was advancing with rapid strides, and it was considered prudent to retrace our steps to the southward, as the blink in the sky denoted that the south water was not far off. To the north was nothing but ice. The next day the ice slacked, and we began to prepare for retreating, excepting the Alexander, of Dundee, Captain Sturrock. Sir Edward promised to assist him into the north water. We gave the expedition cheers, and wished them every success on their mission of relief. They responded, and we parted. We reached the south water, and proceeded to examine every bight or slack part of the ice to the south-ward, in hopes of getting to the west side in time for the whales. We could not get through, so made for Cape Searle. There we got into the west water, and after several stoppages arrived off Agnes' Monument, a little to the northward of the river Clyde. This is a favourite resort for the whales at this time of the year. We fell in with the Alexander, with whom the commander of the expedition had faithfully kept his promise, as a British seaman and gentleman. Every Englishman ought to be proud of the officers of both the army and the navy. I think there is no nation in the world so favoured by having such gallant sons as old England. After we parted company with the Alexander and his friends, they had very little difficulty in getting into the north water. It must be understood that the expedition had steam vessels to tow them along, and there is a wide difference between the using of steam power, and having to drag the ship at a snail's pace. They gave him a tow to-wards the west land. His course lay further south, and their's was for Lancaster Sound. He then parted company with them, and arriving at the land floe, got as far as Agnes' Monument before he sighted any whales. They killed four, and had them lashed alongside. The weather having a nice settled appearance, the master decided to give his men four hours' rest, as they had been up a long time. A gale of wind rose very suddenly, which broke up the land ice, and formed into a pack. The dead whales broke adrift, and all were lost but one, which was a most lamentable piece of ill-luck, and illustrated the old saying, "There's many a slip 'twixt the cup and the lip."

With the land ice broken up, we sailed to the eastward amongst the loose ice, and what is called the middle ice. Sometimes whales are numerous and large in this neighbourhood at this time of the year. Ships have realised good voyages there when the ice has been in good sailing condition, but fortune did not shine upon us. Several whales were seen, but we did not succeed in capturing any. Some ships got one or two. We went a little further south, and killed one off Home Bay. We cruised about a long time, and not seeing a chance of any more, Captain Parker determined to go to Cumberland Gulf, and remain there to the end of the season. He had promised the captain of the McLellan he would go to seek his crew whom he had left the previous year to winter in the Gulf at Kemisuack.

We sailed south, and stayed at different places for a day or two, and sent our boats inshore at different headlands, looking out for whales, until at last we arrived in Cumberland Gulf, and ran up with a fair wind. At Niatlik an American boat came towards us. The face of the American captain lit up with joy. It was a hazardous venture for him to leave fourteen men to winter amongst natives for twelve months. This was the commencement of men wintering there, waiting for the breaking up of the ice, and the whales coming into the Sound. Sometimes the ice breaks up, and leaves a land floe at a distance of twelve or fourteen miles from the place where the parties are wintering. It is a long way to drag the blubber, and tedious to flense the whales. The natives with their dogs and sledges are brought into requisition. In this case the men had been fortunate, the ice having broken up close to their winter quarters, so that it made easier work for them.

When we first sighted the boat, all hands were on the tip-toe of expectation, but as it came alongside, everybody was sorely disappointed. Instead of the American crew, we found they were Esquimaux, in European clothing. There were no lines or harpoons in the boat, but the natives had four or five rifles with them. Of course they were invited on board, but they accepted the invitation rather reluctantly, which made the American captain very dubious. Our first intentions were to keep them on board, and also take possession of the boat. They said the men were all well, had got plenty of whales, and were at Kemisuack, about twenty-five miles distant. They could not, however, give a very clear account of their being in possession of the boat, and we did not trust all they said. Still our captain would not lend himself to any scheme respecting these Esquimaux, as he had such implicit confidence in them. Yet he was not without misgivings. We sailed along until darkness overtook us, and then laid to for the night.

I will here give an account of Cumberland Gulf or Sound. It is about one hundred miles long and sixty miles wide at its broadest part. Each side is studded with small islands and rocks, sunken, and some above water. It has plenty of fiords with anchorages. Rocks extend eight or ten miles from the land, which makes the place most difficult to navigate, and it takes a strong nerve to knock about here in the dark. It was not discovered more than eight years before this date, and of course there were no charts to guide the master of a ship. At various parts whales frequently make their appearance on their way south, especially on the southern side, but as they were mostly on passage, and did not stay in one place, they were seen once or twice, and perhaps no more. If one did get sight of them, they would be miles away. The natives are of a roaming disposition, and do not stay long in one place. In the summer months they go into the interior, hunting deer and salmon fishing in the ponds, some of which are miles inland, and about 1000 feet above the level of the sea. A few years after, in travelling inland, I came to some ponds or very small lakes, and saw many salmon in them. Some might call them trout, yet they had no spots upon them. They were much larger than any trout I had seen, and more like salmon. There is no road to the sea from these ponds, which are surrounded by mountains. It was a question how the fish lived through the winter, as the water is frozen solid to the bottom. Of this matter I do not profess to have any knowledge.

To continue my narrative. At daylight the wind fell light, and we were about ten miles from Kemisuack, so we called all hands to tow towards that place. After four hours' work amongst several sunken rocks, with a boat ahead, we arrived in the harbour and brought up in 14 fathoms. This place or harbour is formed by a lot of small outlying islands, and from the masthead the sea is seen over them. The rise and fall is about forty feet, similar to Niatlick. Still there is no strong tide. The strong southerly current is caused by the melting of the snows from the mountains in the months of June, July, August, and September. The captains landed, but no natives or white men were there. Evidently the place had been lately visited, as the encampment seemed to shew. The natives who were on board were asked about the men, and they said they were at a place called Neubuyan, a name we had not heard before, as the harbour we were in was the furthest any ship had ever been up the Gulf. Our captain was one of the first, if not the first, who had visited this neighbourhood. Things began to look a little mysterious, with uneasy forebodings respecting the fate of the Americans.

D

The following morning the boats were sent away on the look-out for whales, as it would not do for a ship to go hunting about in an unknown country. We had to be patient, and trust to circumstances. The boat I was in being a little further ahead, and more inshore, we espied another pulling towards us, and we gave way to meet it, and found in it some of the men we were in search of. All were dressed as natives. It was indeed a friendly greeting. We told them the news, and they hurried away to the ship. It seems the Esquimaux we had on board had got the loan of the boat for a week or two, so that little affair was accounted for.

The Americans were all quite well, and had not had a day's sickness. They had lived upon Esquimaux food, which is raw frozen seal or walrus flesh. They had been living at Neubuyan, for the ice having broken up in that direction, they deemed it prudent to go there. They had filled all their casks with oil, and killed several whales for whalebone, but had not got any large ones, as they run small in that part of the country at the first breaking up of the ice. Their bread was just done, together with other little luxuries, and they had only about one week's coffee and sugar, having had to feed the natives during the severest part of the winter. They had got so used to the native living that they almost preferred it to civilized food. The company was in command of Mr. Buddington, formerly chief mate of the McLellan, who afterwards served under Captain Hall, in the Polaris expedition. Regarding Mr. Buddington, he wintered several times afterwards in the neighbourhood, and suffered many privations. I corresponded with him for years afterwards, and was many times in his company in Cumberland Gulf, and at Neugumut, a large place close to Frobisher Straits, and always found him to be a most enterprising gentleman. I make these few comments, as there was some reflection cast upon him during the Polaris expedition in after years.

We stayed at Kemisuack a few days longer, and not seeing any whales, we got under weigh, and proceeded to Neubuyan, arriving there before dark. It is a very fine harbour, but has a reef of rocks across the entrance with a very narrow channel between them. There is another road in, but much further away. We brought up, and were immediately boarded by a large number of natives, many of whom we had known before. The Americans and natives were very intimate with each other, for during the whole twelve months they had been on the most friendly terms. All had made this place their head-quarters during the previous six months. The following morning the boats were sent away before daybreak as usual, but before some of them had got their harpoons in the guns, and nicely away from the ship, a gun suddenly went off. "A fall" was called to our great surprise. One of the boats was about a ship's length away from the ship. Something lying very still was seen close by, and on coming nearer it was found to be a whale asleep. The boats always pull as noiselessly as possible, and it had not heard anything until it felt the harpoon in its back. We very soon had it killed. It proved to be rather a small one, but was thankfully received. Our hopes were raised, for we expected we had fallen in with the smaller whales' resting-place, which was Captain Parker's idea. During his life he believed the smaller whales rested somewhere south, but his thoughts were centered about York Bay, Frozen Straits, and Foxe's Channel, up Hudson's Straits. This proved to be the case some years after he died, for many American whalers got full. Many, however, were wrecked there, and suffered severely. After killing the whale, our friends the Americans were kind enough to undertake to flense it, while we went with our boats in search of more. During our stay in this place, the Americans were employed transferring their oil into our casks, and putting their whalebone on board, an agreement being drawn up to the effect that the ship was to have so much freight, and our crew to be paid oil money at the same rate as our own, which satisfied all parties concerned.

During our stay amongst the natives, I was enabled to speak the Esquimaux language, and it was very useful to me whilst sailing to that country. My teachers were two girls, and a boy nick-named Monkey Jack, for what reason I do not know, as he was a good-looking and intelligent lad. One girl went by the name of Hannah, given to her by the Americans on account of her native name being too long. The other was called Kukuya. Several years afterwards the former got married, and came to Hull with a Mr. Bowlby. She and her husband were in the Polaris as interpreters, and were drifted from the ship in a storm, on a large piece of ice, down Davis's Straits, in company with twelve or fourteen men, for a period of, I believe, seven months. It was one of the most remarkable incidents in Arctic annals. The other two, Kukuya and Monkey Jack, were married, and the poor fellow was afterwards killed by a deer, which he had severely wounded, and was going to put an end to. It struck him in the chest with its antlers, and the wound was fatal. I felt very sorry for him, as he always was a great favourite, not only with me, but with all who knew him.

Our captain being a strict disciplinarian, no natives were allowed on board after dark. They could stay until the crew had got their tea, and then they were sent on shore. We used to practice with the bow and arrow until we became quite proficient in its use. Although I was so young my master would take me on shore amongst the skeletons of the whales, and shew me where to put lances in the most vital parts. The skeletons were all far above low water mark, but very little of the flesh was left on their bones. At high water the sharks used to come and feed upon them.

It was very interesting to see such huge skeletons on the land, but more so to see the whole carcase. The sharks are large and numerous, but slow in their movements, and are attracted by the carcases of the whales. When the weather was too boisterous for the boats to go outside, some of us youngsters had the privilege of taking a spare boat to kill the sharks. We struck them with a seal harpoon, hauled them on to the beach, and took out their livers. They seemed to be void of pain, and their livers produced much oil.

Not having seen anything the last few days, we got under weigh for Niatlik, and anchored there before dark. We had all the oil on board, and the ship watered. We brought the Esquimaux with us, their boats and their belongings, which were not numerous. They intended to winter in this place. We sent our boats away as usual, and saw many whales every day, but could not get near them. They were making a rapid passage to the southward. We got ready to leave at any time. The Americans and natives seemed very loth to part, having been so long accustomed to each other's society. The natives were the cleanest I had ever seen. Their dresses were very fancifully decorated with beads, and they had the most beautiful seal skins, which were also very clean. The young female natives wore dried salmon skin covered boots, which shone like silver when the sun was out. They were provided with whale boats, belonging to the Americans, and some harpoons, lances, lines, rifles, ammunition, etc., which could be spared from us. The greatest drawback was provisions. Many natives die through hunger during the winter, as they never provide for the future. When they catch anything they do not seek for more until hunger drives them to it. Their appetites are enormous, but they can go for a long time without food.

The time at last came to take our departure, the bay ice making rapidly, so we hove up our anchor. All the natives came on board to take a last farewell of their American friends and our crew. They were very down-hearted and cried. I believe it would have taken some of our friends very little persuasion to stay another winter, but it was of no use. The natives did not leave us until we were five or six miles away from the harbour, and we gave the poor things three cheers. We saw them a long time lying still in their canoes and boats before they attempted to return. It was five years after this when I saw some of them again. We were six weeks in their company in the years 1857 and 1858, when I was mate of the Emma. The first year of wintering was succeeded by Americans, Scotch, and English, following the same example, and many a poor fellow I have known lies buried upon a small island in the middle of Niatlick harbour.

When the ships winter, they do not carry half their complement of men, but engage the natives, giving them food and clothes in return. It was very detrimental to the habits of the poor things, as their children were not then trained in the use of the bow and arrow or canoe, but trusted to the ships coming. They had got the habit of drinking rum and smoking tobacco, and had contracted other vices. There are no missionaries in this part of the globe. The Moravians were in the Labrador, and the Danes upon the east side of Straits, but not one on the whole of the west side. I could not see it was possible for any to succeed, as the natives were so straggling and always on the move, and I am afraid the poor people were falling back. They were much better off whilst in their wild state. I have been a long time among them, and that is my opinion.

For some days after we had left the Gulf, we encountered heavy weather, and were laid to for three days. The passage home was prolonged by bad weather, and the Americans were very anxious now for home. We arrived at Stromness, and discharged our men. Adverse winds detaining us a day or two, we got a most welcome stock of fresh provisions, which made us enjoy ourselves. With a fair wind we hove up our anchor for home, in the course of three days arrived in the Humber, and soon after docked in the Old or Queen's Dock, and found all our friends and relatives well, which made us forget the trials and vicissitudes of another Arctic voyage.

CHAPTER V.

THIS voyage commenced early in March. Having all
prepared, we left Hull with a fine fair wind, and in three
days arrived at Stromness in the Orkney Islands. We only
stayed here two days to ship our complement of men, and
again got under weigh, and proceeded to the westward
through Hoy Sound. This time I was promoted to the rank
of boat-steerer. We took all our harpooners from Hull.
Formerly part of our principal officers were from Scotland.
One or two of the present officers were old men in the trade,
and were very slow compared to those of the present day.
They seemed to work in one groove, and could not on any
consideration get out of it. I was appointed boatsteerer to
a rough but nice man of the old school. You could not
choose your boat or watch. They are all divided into three.
The plan of allotting or choosing was thus :—When everything was stowed, and lashed ready for sea, the crew were
all called aft (the boat-steerers first), and placed in six rows.
Next came the line managers and the other members of the
crew. When the rows were filled up, the harpooners gave
their pocket knives to a boy who had not been to sea before.
He placed a knife at the foot of each row, and those to
whom the knife belonged became the harpooner's boat's
crew for the voyage.

The captain's boat consisted of the cook, carpenter,
cooper, cabin boy, or steward, and one or two spare hands.
The boats were allotted in a similar way, so there could not

be any favour shewn, or a better boat picked by anyone.
Three cheers were given, grog served out, and the watch
then set. During the passage outward, everybody was preparing for the fishing, weather permitting, harpoons, knives,
and other instruments were sharpened and cleaned. We
had a remarkably fine passage, and arrived off Fortune Bay,
in the island of Disco, about the 12th April. There we
sighted a whale, but did not succeed in capturing it. We
went to the offing, with Riff Koll about sixty miles and
Disco about sixty-five or seventy miles distant, and the ice
was in a very favourable condition, being composed of
loose floes and sconces. We were soon employed amongst
whales, as this place was noted for them at this time of the
year, when the ships could get into the neighbourhood.
There was a very old saying, "With Riff Koll Hill and
Disco dipping there you will see the whale fish skipping,"
which saying was verified. Our boats had no sooner been
lowered than we succeeded in getting fast, and had the
whale killed in about two hours. The following day the
brig Anne, of Hull, hove in sight. She belonged to the
same owner. She at once fastened to one, then to another,
when fortune appeared to desert us and go over to the
Anne. No sooner did we strike a whale, than she followed
suit, theirs was killed immediately, whereas ours took many
hours. His crew had all the play, and ours all the trouble.
When we struck our third whale, he immediately struck
another. Ours took out four boats' lines and capsized a
boat, and was with great difficulty captured after twenty
hours' battle. The Anne got two during that time, and
flensed before ours was dead. A strong wind came on and
began to close the ice. We then went to the northward, off
Disco Bay. The land was about twenty miles distant, the
ice here was formed into a pack; the wind blowing on to it
made a very nasty short sea at the edge. A few whales
were seen, and we were soon in pursuit. One was struck,

and she led us a nice dance. We put three harpoons in, then she went into the pack. As the boats were entering the ice, one was capsized, and another severely stove. The men belonging to the boat which sank had to hold on to each other on the ice, as no boat could render them any assistance on account of the sea. The ship took the other boats on board. This was the time to shew what a smart little craft the Truelove was. The captain was carrying on a press of canvas. He ran the ship into the ice, threw lines to the men, got them on board, hauled to the wind, and the little barque came out of the ice splendidly. The other two boats having got about three ships' length inside of the pack edge, did not feel so much of the swell, but it was a most critical time, as the ice was rapidly drifting past several large icebergs inside of the pack. We made preparation in case the boats were likely to come in contact with them. The whale at this critical moment sought the open, although this is not frequently the case.

We were soon dragged into the open water again, and the ship was hovering well to windward, ready to drop the boats in case the whale should rise. She did not do so, and we came to the conclusion that she was dead at the bottom. We took the lines to the capstan, and began to heave in. All hands was sure she was dead, especially when we detected something white not far under water. We gave three cheers, but, alas, we were disappointed, as it proved to be the sunken boat which had got foul of the other boats' lines. It must have been dragged along the bottom for hours. The harpoon gun, which is fixed in the bow of the boat, was polished and choked with mud, of which the bottom is composed, at the depth of 120 fathoms. In an hour's time we had got the lines in but no whale, the harpoons having drawn. The weather now was fine, but our disappointment was great, after twenty-four hours toil, in addition to being wet and our clothes frozen, still we did not despair, but got our boats again prepared, and the watch set ready to begin again.

The following day we got fast to another whale, with hopes of better luck, but this one also gave us much trouble, as she took the pack, and no boats could follow her, the ice being so tight. We had to bridle on, that is, a boat on each side of the fast boat bent his lines on to the fast boat's lines, so there is the strength of three lines instead of one. This is a last resource, for it must either break the line or draw the harpoon. Hands were sent over the ice with lances, in case the whale made her appearance in a hole of water. In a short time the captain got sight of her, being attracted by the malemauks hovering around. Some of the men put a lance into her, which soon made her spout blood, and shortly after she was killed. We took the lines to the ship and hove the whale alongside, but this was not done without a lot of trouble and anxiety. We were afraid of the harpoon drawing, owing to there being a strain upon it, and having to drag the fish from under the ice for such a distance. This made our fourth whale. The Anne had got six with very little trouble. We saw some ships to the southward, and the time drawing on we plied to the northward, towards Black Hook or N.E. Bay, and there fell in with more fish. We got fast to a very large one, which had a young one with her. She made for the ice, and one of the old harpooners was so near that his boat touched her back. He fired his harpoon but missed. He was so confused that he never attempted to use the hand harpoon. This blunder had given the whale breathing time and fresh strength, and before any of the other boats had time to come near her, she had taken the ice, which was not so tightly packed, but allowed her space to rise for breath. Still the boats could not go amongst it. The whale took out fifteen lines in succession, and the other boats not being able to get near to the fast one, there was great danger of losing the whole of the lines. However, we came to a large piece of ice where we secured the end of the lines before it had run out of the boat. The whale dragged the piece of ice along at a fast walking pace, then suddenly stopped. We knew something had given way, and we began to haul in the lines. This is no light work for three boats' crews. It represents 1800 fathoms or 10,800 feet of line, and perhaps two-thirds are laid on the bottom. It made a long, heavy drag which took us many hours to accomplish. If we had got a second harpoon in when there was a chance, it would have saved us all our labour. Such a mishap was very disheartening. The next day we were fortunate in securing another one, but not without a great deal of trouble.

The Anne by this time had been exceedingly fortunate, and the fishing on the east side being over, her captain determined to make the best of his way home, and not risk the dangers of Melville Bay. The other ships with ours proceeded north through the Bay, but not without its trials, and across Baffin's Bay to Lancaster Sound. We were late for Pond's Bay. We found an opening close to the land on the south side, and saw several whales. We sent all the boats, and followed with the ship. One of the boats got fast to one, and although not a very large one, she took out twenty-four lines without stopping. This is an exceptional case, as she went under the floe. In this part they do not take out more than six or seven lines before they come to the ice edge to breathe. The ship was by this time made fast to the floe, close to the boats, and the lines not running out, we commenced to haul them in with the capstan, which was fifteen hours' dreary work, and then found the harpoon had broken. We had hauled in 17,280 feet of line, and had nothing to repay us for our labour. We gradually sought south, calling at Coutt's Inlet, and Scott's Inlet. We fastened to several more fish, and lost them with harpoons drawing or lines breaking, one with

lines running foul, capsizing the boat. We eventually
secured nine whales, making about 110 tons of oil, and
seven tons of whalebone. We had lost during the season
ten large whales, without counting several misses with the
gun and hand harpoon.

We gradually drifted down the country with the ice,
having our boats away frequently, but not succeeding in
getting any more fish. We intended going to Cumberland
Sound for the remaining part of the season, and got as far
as Cape Dyer, nearly all the ships being in our company,
among numerous icebergs, but clear of the other ice, when
we met with a heavy gale. It blew some of the sails away
that were stowed, and split others. Our only remaining
sails were a close-reefed main topsail, fore staysail, and
main trysail. A very heavy sea was running, and thick
snow falling. To add to our misery, darkness was coming
on. Never before or since have I experienced such a night,
and I hope I never shall again, or anyone else. All hands
were up, looking out and managing the ship, which took
their utmost skill to do. We succeeded in keeping clear of
many bergs until about midnight, when in the act of
wearing the ship for a very large iceberg, we struck a piece
of one, which stove in our port bow, and broke twelve
timbers. We immediately got swabs, with which the
vessels are well provided, and an old sail over the damaged
part. The carpenter put some shores inside, and with a
good force at the pumps we kept the water under until daylight. The gale then moderated a little, and we got the bow
made more secure. At noon we set the watch ; there was a
very high sea running, with misty weather. The ship was
laid to. Suddenly there was a cry of " a man overboard."
Everybody rushed on deck instantly. It was one of the
harpooners, named McKenzie, who had gone over the bows
to see what extent of damage was done. Being muffled up
and wearing oilskins and sea boots, also a pair of Esquimaux

mittens, which are very slippery, it is supposed he had missed his hold, and that a projecting piece of the plank had struck him on the head. His face was under water when he floated close by. I threw him an oar, and called him by name, but he never once lifted his head. A boat was speedily lowered, and although a high sea was running, they pulled in the direction he was last seen, but had to return without him. It was a great risk for the men in the boat, but that was not thought of at the time. This cast a deep gloom upon the whole crew, as poor McKenzie was a thorough seaman, and beloved by us all. The following day was beautifully clear and frosty. The wind had changed from S.W. to N.W., so we proceeded towards Holsteinberg, a large Danish settlement upon the east side, and shortly after came to anchor for repairs.

The next day the exploring steam ship Pheonix, Captain Inglefield, came into the harbour. He ascertained our condition and sent his carpenters on board to help us to repair the damage. He also wished to know if there was any danger of the other vessels, but our captain thought not. If there had been, he would have left his carpenters and steamed across the Straits to aid them. Captain Inglefield sent us some preserved meat and vegetables during the time the repairs were going on. We got watered, boats stowed, and everything prepared for our homeward passage. In three days all was completed. The Pheonix towed us out of the harbour, and sixty miles to sea, leaving us with three cheers. Now this was another good action done by one of our noble naval officers. They are not only gentlemen, but thorough seamen. In the course of three weeks we anchored in Stromness, discharged our men, and the following day got under weigh, and in due time arrived home. We found all well, which was a great relief to my mind after such a voyage as we had experienced.

CHAPTER VI.

THE LAST VOYAGE—CHANGE OF MASTER—CAPTURING POLAR BEARS.

NOW comes the sixth and last voyage of my apprenticeship, and with it a change of captain. My old master left the ship, and was succeeded by Captain W. Wells, late of the Anne. We sailed as usual with great expectations, and were favoured with a very fair passage to Lerwick, where, as formerly, we shipped our complement of men, and proceeded to Davis's Straits. We had a fair passage out, and made the first ice off Goodhaal, on the east side, in Lat. 64° N. By the first ice, I mean that which consists of streams of ice or pack ice. Icebergs are so frequently fallen in with that they are not mentioned. We were about twenty miles from the land, and threaded our way to northward, among loose and streams of ice, until we reached Whale Fish Island, in S.E. Bay. It is one of a small group of low islands, and has a settlement, the governor being a Dane. The natives came off, and said that many whales had been seen the week before. We cruised about, sometimes inshore and at others in the offing. At last we struck a whale and soon got her killed. We plied about for some time without success, but had our boats away frequently, not seeing anything for several days. We proceeded further north, and with great difficulty reached N.E. Bay, having had to go inside Hare Island. The quantity of ice in the offing leaving very little of what is called the Black Hook water, we were compelled to go up the Bay, and there we got another whale. This Bay had not been explored, and the

Esquimaux told us there was no end to it.   Of course the
numerous fiords, glaciers, and islands held out no induce-
ment for them to go.   Their hunting and fishing grounds
were nearer home.   On the north side great icebergs were
very numerous.

In course of time we again proceeded north, but were
longer on account of the ice lying so near to the land.
Whilst we were made fast to an iceberg off the settlement of
Upernavik, a volunteer boat's crew was formed to go to
the interior to get duck eggs.   We returned with thirty
dozen, and a quantity of eider ducks, having been away
twenty hours.   During our trip in the boat we witnessed an
interesting incident.   A massive iceberg broke away from
the glacier with a tremendous crash.   It roared and cracked
before it fell like artillery firing.   The fall caused a heavy
swell to roll upon the small islands in the neighbourhood for
two or three miles.   These little expeditions were called
pleasure excursions.   We saw fresh places, and it was a
change from being cooped up on board ship.   When a
volunteer boat's crew was called, there were four times the
number of men responded to it.   By tracking, and principally
towing, we got as far as the Devil's Thumb, but no further.
Melville Bay had not broken up, and there was no appearance of water, excepting the thin line to the southward.
As the season was growing late for getting to Pond's Bay, it
was resolved to retrace our way back.   The innumerable
heavy icebergs aground, stretching far to the westward, upheld the large floes, and it was an impossibility for us to
seek a passage through the Bay.   We went south again.
It was surprising to see the alteration in the appearance
of the ice in so short a time.   What were large floes and
apparently solid ice a month before, was now clear water.
In these northern climes the greatest drawback consisted of
long dense fogs, which made us very anxious, as they were
at all times dangerous.   In coming south we kept along the

ice edge, and tried every bight as usual until we reached the tail end of the ice, off Cape Walsingham, on the south side of Exeter Sound.   The ice was packed upon the west coast, so we again proceeded north, about 70 miles, and entered a deep bight which we had passed during a dense fog.   It would have been much better to have stayed until it had cleared away had we known.   Through this deep bight we had no trouble in getting into the west water, the ice forming the south part of the bight had separated from the main body.   Some of the ships had got across several days before.   We ran along the west land in clear water with a fine S.W. wind until we came to Home Bay, but saw no whales.   We went off amongst the slack ice in the offing.   Several whales were seen, and boats sent in pursuit, but without success.   We shot several bears, and captured two fine ones alive.   We used a lasso to take them.   The rope is rove through the boat's ring in the stem, and thrown over the bear's head.   When it is over, you must be very smart to take in the slack, but once the animal is drawn tight up to the ring of the boat, it is secure, and cannot do any mischief unless any unforeseen occurrence takes place.   It is rather a dangerous performance, particularly if the boat's crew are not quick, or if there is any clumsiness on the part of the harpooner or boatsteerer. The bears were got on board, and made securely fast to ring bolts in the deck, with ropes round each foot and neck until the cooper prepared a cask with iron grating at one end.   I heard tell of a harpooner, the year before, forgetting to reeve the rope through the ring of the boat, and when the noose was round the bear's neck, he climbed into the boat.   Happily for the crew the animal jumped out again. They ran to the after part of the boat, all the weapons were forward.   It was always the best and safest plan to have a gun and lance ready.   In this case the bear was shot.

Cruising about for several days, and seeing no whales

E

or any prospect of getting to the northward, we were
obliged to seek inshore, and fetched in by Cape Broughton.
The pack ice was drifting along the land, but no distance
off, so all the ships sent their boats inshore at this place to
see if there was anything to pick up. We went to Cape
Searle, and there laid at anchor about a week, sending the
boats away every day. The weather during this time was
beautiful. The sky was clear, and there was little wind. It
was a pleasure to be away in the boats. We were obliged to
get under weigh and go outside, as the ice was coming down
rapidly. We again went to the north of Merchant's Bay,
and sent our boats inshore. There are two small islands or
rocks, called by the men Terrification Islands, but by whom
and for what reason I do not know. One day boats were
sent away from all the vessels, and men landed on these
islands. They took notice that the bay was fast filling up
with the ice coming down along the mainland. Those who
knew the place well, said how dangerous it was to stay
longer on account of the ice surrounding the islands—the
currents eddying round would in a short time prevent their
return. They warned three boats' crews belonging to the
Eclipse, of Peterhead, of the consequences of delay. The
officers belonging to that ship, being strangers in Davis's
Straits, took no heed of the friendly warning, but pulled
further amongst the loose ice towards the mainland, thinking
there might be a better chance of securing a whale there
than at the outside. Their ideas were correct if there had
been no ice to contend with, but what use was it to get
trapped when it could be avoided, and lose both chances.
When they desired to return they were unable to do so, as
the ice had completely encircled them. The other boats
returned to their respective ships. The following morning
nothing was seen of the three boats. The wind came to
blow strong on the land, and there seemed to be little hopes
of ever seeing them again, the ships having to work off from

the ice. The third day it became fine and clear, and they went close to the pack again. The missing boats were sighted, and some men were seen travelling towards the ice edge. A boat was sent, but was unable to reach them on account of the broken condition of the ice, and they could not walk upon it, consequently they had to return to their boats.

Next morning the Heroine, of Dundee, being the first at the ice edge, and a light air of wind coming off the ice and blowing the light stuff away, succeeded in getting the men on board in a most deplorable state. They were put to bed and carefully attended to. After a few hours' comfortable rest they fell in with their own vessel, and were put on board. A few days afterwards the Eclipse found his boats drifting out of the harbour of Durban. This is the case before mentioned. I think it was in this year that the Eclipse, Capt. Gray, came from Greenland to Davis's Straits, crossing over to Pond's Bay in clear water. He found Pond's Bay clear of ice, and ran through it into a large Sound, since called Eclipse Sound. He was the first who had been so far that way, and that solved the question of whether there was a direct passage through to Prince Regent's Inlet and Lancaster Sound. It was through Eclipse Sound that Captain McClintock got his passage so well into Prince Regent's Inlet, in search of Sir J. Franklin, in 1858. Captain Gray found plenty of whales in this Sound, but the ice was so rotten and full of holes that he could not get near them. Otherwise he would have done well, and none of the Davis's Straits men would have known he had been to the country.

The ice began to come down rapidly, and gradually drove us further south. It was thought best to make our way to Cumberland Sound. We therefore anchored in Niatlick harbour and got a whale, our boats going away as usual. A few of our old Esquimaux friends were there, but many were dead. The majority were further up the Gulf. It was in this place our mate died two years afterwards, and was buried upon a small island lying in the middle of the harbour. I saw his grave when I was there as mate of the Emma. As I stood by it I thought of him in the prime of life. He was a fine, strong, healthy man only a short time previously. We stayed here till it was not safe to remain any longer. Not knowing the state of the ice outside, it was deemed prudent to make the best of our way homeward, so we left with a fine, fair wind, and a full moon to light us out of the country. After a favourable passage we reached Lerwick, received our letters, newspapers, and news of the Crimean War. They are most anxiously looked for by us, as we receive none from leaving the Shetland or Orkney Islands until our return. The Arctic Regions does not boast of a postal delivery, so any news is eagerly devoured. We soon arrived once more at Hull, and found all at home quite well.

CHAPTER VII.

CONCLUSION—ADVICE TO APPRENTICES.

NOW if any youth, who is intending going to sea, should read this rough sketch of the life of an apprentice, I would advise him to be very careful how he enters upon his duties. He should be civil to everybody and dutiful to his officers, doing his best to gain their good-will by performing what he is told, cheerfully. When he is set to do anything, do it quickly with a good grace. Nobody gains ill-will so soon as a sulky, grumbling boy. I will vouchsafe to say at the end of a long voyage a civil boy will be respected. Do not listen to the yarns of some men. When they wish you to stay, leave at once, and begin some trifling job, also improve your mind with reading, and your spare time in learning navigation. When the men see you are superior in education to them, they will treat you with respect. If a poor fellow cannot write, proffer to write his letters for him. It will cost nothing, and he will send a letter to his friends, otherwise he would neglect doing so, and I can assure you that he will befriend you in some way or other. Help those who are not so well educated as yourself, and do not taunt them because they are not so, although there are not so many now as formerly who cannot write.

These few remarks I hope will not deter any youth from going to sea. The times are much altered since I first began. Boys are well looked after, and their comfort studied. As regards the hardships of the Arctic regions, steam has superseded sailing vessels, the work is easier, and all the hours which used to be spent in towing and tracking, belong to the past. I have gone through Melville Bay with steam in twenty-four hours, without losing a moment's rest, whereas I have been six weeks going through with a sailing ship, and no better prospect in view. So things are more easily done than formerly.

Here ends my apprenticeship. I became master of the Truelove in the year 1861, to Davis's Straits. Like mortals, ships have a termination. The old Truelove—all honour to her memory—no longer exists. She is broken up. Her stout timbers suffered many a tight squeeze in the great ice fields of the north. The good ship went through all these without being seriously endangered, and with the close of the Hull trade to Davis's Straits her occupation was gone. Peace to her ashes.

CHAPTER VIII.

A CAPTAIN'S ANXIETIES AND CARES—FIRST VOYAGE AS AN
HARPOONER—TALES OF THE PRESS-GANG DURING THE
WAR WITH FRANCE.

I HAVE been requested by many of my friends to give a
further account of my adventures in that inhospitable
country which came under my observation after my
apprenticeship.

It would not have been wisdom to relate my experience
as an officer in that book which contains the history of my
boyhood days. Some people may think that a person's
mind cannot recall events dating so far back, but I will
explain how that is possible.

During my apprenticeship I was obliged to keep a log,
and was taught to take notes of particular events which
frequently occur in that cold region.

My master from the first took great pains in teaching me
the appearance and marks of the land which, I have before
stated, as there are no charts to guide you among the
numerous sunken rocks which abound on both sides of
Davis's Straits, especially on the west side, in the neighbourhood of Cumberland Gulf and Frobisher Straits, of
which places I had some experience as mate of the barque
Emma, in 1857 and 1858 and afterwards.

A person who admires scenery has many opportunities of
observing it from the crow's nest of a whaler, especially
when the ship is not on whaling grounds. There his mind
is centred upon his occupation, and his eyes ache with the

constant use of the spyglass ; yet there are times afforded
for study, and grand sights to be seen from that elevated
place. You can picture yourself in the crow's nest, which is
fitted with a seat and a step to rest your feet upon, and has
also a weather board to protect you from the wind. Now
supposing the ship is made fast to an iceberg, close to the
land, waiting for an opening of the ice, and the time midnight, weather clear and calm, the sun shining beautifully.
In the offing nothing is to be seen but close packed ice and
numerous icebergs of all shapes and sizes. The land close
to is bare and rocky, with clear sparkling water running
down the steep valley from which we have just watered.
All is so quiet that the stillness appears oppressive, yet the
beauties of nature are so varied as to wear off that quietness.
The land in places is very high, and there are glaciers in
the distance with the sun shining upon them, which gives an
indescribable grandeur to their appearance. Then is the
time to see and study nature in her wild and beautiful state.
Sometimes a very sudden change comes over the scene,
perhaps four hours hence. In the south, clouds are
travelling fast to the northward, and the sun begins to look
greasy. The wind comes in light gusts, and then follows a
gale. All hands are called to get the ship into a more
sheltered position if possible. It is now blowing a severe
gale, with thick snow, although the month is June. The
berg keeps the pressure off the ship. The ice wraps round
it and lays great strain upon the warps, sometimes breaking
them, or the anchors jump out of the holes. This is a most
critical time until the ice becomes tightly packed behind, or
the gale abates. This shews the two extremes which so
often occur, and proves that a master's life is not a bed of
roses. When he retires to rest he may be called up at any
moment, so his mind is constantly agitated, asleep or awake,
especially is this the case with the man who has not been
successful. The only pleasant time he enjoys is on the

passage home after a good voyage, and with his crew in high spirits.

The first voyage I made after being out of my apprenticeship was as loose harpooner of the brig Anne, and it proved to be a most unsuccessful one. Our first trouble was the loss of a young Shetlandman, who died of consumption, and was buried at sea with the usual solemn formalities. The following day a gale of wind carried away our maintopmast, which lost us the run of a fine wind up the Straits, being thus detained twenty-four hours to the eastward of Cape Farewell, and the wind changing we had to beat about three weeks in order to double it, and thereby losing the best part of the east side fishery. Ships generally gave the Cape a wide berth, sometimes ninety or one hundred miles, on account of the heavy ice which drifts from the west coast of Greenland and the east coast of Davis's Straits, and meet, forming a pack or streams sometimes extending a long way south, also numerous icebergs which ground off there, making it dangerous to approach nearer in the early part of the year. Icebergs are beautiful when the sun shines upon them, and light up their massive forms. We see them imbued with all the colours of the rainbow, and their appearance varies to a most surprising extent. Some are wall-shaped, with flat tops, others rounded, and many-pinnacled like church spires and Turkish mosques. Some people think that icebergs form in the sea, but it is not so ; they break off in massive pieces from the foot of the glaciers which are so numerous on the east side of Davis's Straits and Baffin's Bay. Those glaciers I consider to be one of the greatest wonders of the world. Terror and beauty combine in this place of desolation. Many a good ship has never been heard of since crossing the Banks of Newfoundland, but if those bergs could speak they would solve the mystery.

Of this voyage it is needless to make much comment. On

our arrival off Disco we encountered much ice, but in the course of a few days the wind came off the land and cleared a passage to Goodhavn. The master went on shore to see if he could get a spar for a spare topmast, as he knew that the wreck of the Rose, belonging to Hull, lay on the beach in the harbour. She was severely stove to the N.W. of Horse Head the previous year, but they managed to get her to this place. Fortunately, there was one which suited us, and, having made arrangements with the Governor, we dug it out of the ice and towed it to the ship. Many whales had been seen before we arrived. If we had not lost our top-mast, we might have been here in time to have caught some, shewing what a little may change the whole aspect of affairs. We cruised about for a few days, but no whales were seen ; then gradually sought our way to the northward towards Swarte Huk or Black Hook, which forms the north part of N.E. Bay or Omenak Fiord, in the offing of which extends fine whaling grounds in the month of May, and is the last on the east side of Davis's Straits. We had great difficulty in getting through the Malygat Straits, which separate Hare Island from the Island of Disco. This detention caused us to be too late for the fishery, but we could not get further than Upernavik. The ice appeared not to have broken up to the northward, so we made fast to an iceberg near a large glacier, where we stayed some time in hopes of getting north, but finding no probability of doing so, we returned south and tried every bight, but could not get nearer the west land within forty miles. It was a most weary and trying time, nothing but a solid mass of ice to the westward was seen. In one deep bight we observed from the crow's nest a spar standing upright in the floe about four miles distant. A party of us travelled to it, and found the ice of very great thickness, and the surface had three feet of water upon it, and to all appearance a vessel was underneath. The ice all around was of some years' accumulation, and called by us

Sound ice. Old hands say this kind of ice only breaks up
once in seven years. It generally drives down the middle
of the Straits, and comes from Lancaster and Jones' Sounds.
Coming further south, we sighted a vessel in the ice a little
south of Cape Dyer. At first we thought it was one of our
unfortunate friends who was beset. Later on we heard it
was the Resolute, belonging to Sir Ed. Belcher's expedition,
which was afterwards picked up by Captain Buddington, an
American whaler, who was trying to get into Cumberland
Gulf. We cruised about in all directions hoping to find
whales, but there was no prospect of getting to the west
side this year, and the weather becoming boisterous we bore
up for home in the first week of October, and after a long
passage arrived at Lerwick. One of the boatsteerers
belonging to that place had been laid in his hammock for
six weeks with a broken thigh. None of us knew what was
really the matter, as we did not carry a doctor. We had
rubbed it with linament, thinking it was a flesh rent, but the
bone united falsely, and was obliged to be broken again
when he landed. The poor fellow was a cripple for the
remainder of his days.

Our arrival in the Humber recalled to mind many stories
told to me by old men respecting the press-gang during the
war with Napoleon, some of which I will relate. In those
troublesome times, harpooners, boatsteerers, and line
managers were exempt from being pressed if they could get
to the Custom House and receive their protection tickets,
as they were called. Many a ruse was enacted to reach that
place. One vessel on arrival at Hull was waited upon by the
press-gang. The captain, whose name was Sadler, ordered
his men to arm themselves with the sharp and dangerous
weapons which are used in flensing whales, and marched
them to the Custom House to receive their protection
tickets, thereby enabling them to go safely home to their
friends. On another occasion the press-gang attempted to

board a ship in Hull Roads, but were told if they did so that violence would be resorted to. The warning was not heeded, and, in consequence, the lieutenant who led the party lost his life in the scrimmage. He was buried on the east side of Drypool Churchyard, and the inscription on the gravestone states he was inhumanly murdered. The press-gang was chiefly composed of the lowest class of seafaring men utterly void of any tender feeling. It is no wonder that men who had been months in such a desolate region should resent such treatment. Some captains (weather permitting) would land the greater part of their men at Tunstal or Easington on the Yorkshire coast, or inside of Spurn, and the poor fellows, if they escaped capture, would have to steal to their homes like thieves in the night. It is a great blessing that England has dispensed with such resources to obtain men for the Navy. Other devices were resorted to in order to frustrate such unwelcome visitors. A revenue cutter hove in sight off Flambro' Head when Captain Scoresby was returning home with a full ship. When he saw it in the distance, he let four or five feet of water into the hold through a large brass tap which some whalers had in their counters on purpose to fill their casks for ballast. This was kept running, so that the pumps could not gain upon it, and when the officer boarded the ship he was told she made so much water that the crew would not be able to keep her afloat if he took any away. The officer sounded the pumps, and was satisfied in finding when they stopped pumping the water rose in the hold. He took his departure. The tap was at once turned off, and the water pumped out. This clever trick saved his men from being forced on board His Majesty's ships. It was also related that some men were smuggled into the Custom House dressed as women. Others were taken there in casks in order to avoid the press-gang. Once inside the gates they could not meddle with them. Another incident took place during the war. Two

French corvettes cruised in Davis's Straits at the old S.W.
lat. 62½° and 63° N.   This place was a favourite resort of
the whalers at that time, but a most dangerous place with
the wind blowing on the pack.   They captured two or three
vessels and burnt them, taking their crews prisoners.   Captain Sadler, whom I have previously mentioned, saw one of
the Frenchmen at a distance, and a less experienced man
would have taken him for a whaler.   Having watched the
strange vessel for a time, he remarked the whaling gear was
on the starboard side instead of the port side.   Captain
Sadler at once headed his ship towards the pack, crowded
all sail, and ran into the ice.   A high sea was running, but
he preferred the risk of losing his ship in the ice to being
taken prisoner by the French.   The enemy dare not follow
him, and when he got a few miles inside the pack, he lay
comparatively safe until his pursuer disappeared.   The risk
was very great, for many a noble ship has been lost with all
hands in the Greenland pack.

CHAPTER IX.

OLD TRUELOVE — GREENLAND SEALING — MELVILLE BAY
SQUEEZES — LOSS OF THE PRINCESS CHARLOTTE —
WHALING AND DANGERS OF WHALING.

IN 1856 I engaged as harpooner of my old ship Truelove,
and our destination was Greenland. If not successful
there we were to proceed to Davis's Straits. We fitted out in
the corner of the Old, or now called Queen's Dock. The
crow's nest was aloft with the effigy of a man on the look
out. This attracted the attention of passers by. There
was huge quarters of beef in the tops, and in every respect
well fitted out. Everyone on board were in good hopes of
a prosperous voyage, and the vessel was viewed by hundreds
of passers by while lying ready to sail with the first spring
tide. She was but a small barque, yet drew about 16 feet.
When the time arrived for our departure, we were towed
down the old harbour and into the river amid hearty cheers
from the spectators. Arriving at Shetland, and obtaining
our complement of men, forty-five in number, we sailed
for the sealing grounds at the west ice of Greenland, in
company of several more ships. This season proved very
rough and boisterous, the winds prevailing from the eastward, closing the pack so tight that we could not force our
way into it to look for seals. A few of the bladder-nose or
hooded seals were shot. These animals, when attacked,
inflate a loose skin above their nose, like a bladder, which
protects the head from the blow of a club. The best
weapons are the sharp axe or rifle. One day I went with
my boat's crew to kill two of them, which were on a large

piece of ice. It was blowing hard at the time I landed. In
jumping out of the boat it drifted away before the man had
time to secure it, leaving me alone on the ice with only one
club. When I struck at the head of the male seal he seized
the club with his mouth and wrenched it out of my hands.
In the meantime two more men landed with their clubs, and
another one for me. But the furious animal evaded our
blows. We found our weapons useless, so returned on board
for a rifle, and quickly despatched them, the crew on board
having a hearty good laugh to see us retreat. The male is
a most courageous fellow, and will not leave his partner,
but will stay by her until he is killed. Some have a severe
fight with a bear before they give in. Having tried in vain
to fall in with the seals, our thoughts turned to the prospect
of whaling in Davis's Straits, but a sudden heavy gale
arose from the eastward, and being close to the pack we
were forced to enter it, with a heavy sea running. The
strong little ship got a short distance inside and came in
contact with a heavy piece of ice which her weight would
not force aside. This made her broach to and lie
thumping and bumping for several hours. We had to cut
up several warps and make them into small coils to protect
her sides from the sharp corners of the ice. This work
required the attention of all hands during the whole time.
At daylight the following day the gale moderated, and the
swell went down. It took us three days to get into clear
water, and according to our instructions proceeded for
Davis's Straits with a fair wind. Our garland was put up on
the 1st of May, honouring that auspicious occasion with a
similar ceremony to that observed in crossing the line,
particulars of which will be found in a subsequent chapter.
Nothing particular occurred until we arrived off Black
Hook, on the north side of N.E. Bay. There we fell in
with the Davis's Straits fleet, bound north towards Melville
Bay, and spoke the Emma, of Hull. They had wintered

near to Goodhavn, in the island of Disco. Captain
Parker not being able to get into Cumberland Gulf on
account of the vast amount of ice which laid off that place,
as the preceding chapter relates. The crew of the Emma
had killed two whales in the month of February, and seen
great numbers, but owing to the young ice forming it was
difficult to get near them. They stated that the month of
February was much milder than the beginning of May.
All the ships now began to make the best of their way
north, with the usual routine of tracking and towing when it
was calm weather, until the north water was sighted off Cape
York. A strong wind sprang up from the S.W., which
quickly closed the ice. All sail was carried on by the
ships to reach the water before the ice came upon us.
Several succeeded in getting into it, and were closely
followed by the Heroine and Princess Charlotte, both
belonging to Dundee. A large sconce lay between the two
floes, leaving a ship's breadth clear on either side. The
Heroine took the weather side, and the Princess Charlotte
the lee side, which could not be avoided. The ice was
closing so rapidly that it squeezed the former into clear
water and jammed the latter between the floes. In less
than ten minutes the ice went through her, and the masts
were laid on it. The crew had scarcely time to save their
clothes, and the master only time enough to get down from
the crow's nest. She had all sail set when her masts fell,
and her cargo consisted of five whales, seventy-five tons of
oil, and four and a half tons of whalebone.

The other vessels, which were close astern, were soon
fast, then came the heavy pressures upon them, and a scene
of the greatest confusion, what with the roaring of the gale
and the crashing of the ice, the crews of the ships launching
their boats away into safety, getting their clothes and
provisions on the ice in case their vessels should share the
same fate as the Princess Charlotte. Imagine the com-

motion, there was eight ships within a radius of half a mile, and all were laid in different positions as the ice forced them, there was no time for sawing docks or any room to move, so we had to trust to Providence to protect us and help us out of our difficulties. To make it more uncomfortable, it came to snow and sleet that you could not see a ship's length, and blew a perfect gale, the ice now and again giving an extra squeeze, making the ships groan and jump ; it was a most anxious time, as every pressure we expected would be the last of us.

This continued for fourteen hours, and then gradually the gale died away, leaving a most desolate scene behind, and kept us tightly packed four days. As no water was to be seen from the crow's nest, we took the opportunity of getting the provisions on board and making ready when the ice slacked. The different boats' crews of the wrecked ship went on board of the other vessels, the second mate with his men came to us. I am sorry to relate the conduct of the men belonging to one boat. They would not go on board of any ship, but presented one of the most beastly pictures I ever beheld. Some of them had evidently been concerned in a scuffle, their clothes were torn, one or two with black and inflamed eyes, staring at us who had come to rescue a man belonging to their own vessel who had imprudently gone to persuade them to give up their drinking and go on board the ship which was allotted them. They had erected the boat's sail tent-like for shelter. Their clothes were strewn about, and the lances were laid on the ice, ready to act on the offensive or defensive should anyone attempt to meddle with or advise them ; also part of the whale lines were hacked and chopped in pieces for fuel, whereas if they had taken them on board of the vessel which was to be their future home it would have partly recompensed the owners of that ship, but no, they deliberately let the remainder go overboard and sink to the bottom rather than anybody should benefit the

F

least by them. Such selfish beings are not often found. Their portion of rum, which had been grappled from the wreck when the ice had been cleared off the deck, was distributed to each boat's crew. They preferred to drink it and make beasts of themselves rather than take it on board another ship, although it would have been kept for their sole use. When the ice began to slack they went on board of the vessel which was allotted to them ; all the men belonging to the Princess Charlotte were put upon full pay, excepting these.

The scene now presented a far different aspect to what it did four days ago. All was animation, getting the ships into position as the ice slacked. We then began to track and tow to the northward in a narrow channel between the floes, but no open water was to be seen from the mast-head, the gale having filled it up for miles beyond. Three days elapsed before we reached open water. The change was hailed with delight, and the usual three cheers were given.

With a fine breeze all sail was set, and the ship was headed towards the West Land, which we made on the north side of Pond's Bay. Here we fell in with the other vessels which had succeeded in getting clear of the ice before it closed upon us off Cape York. Some had got several whales, which made the prospect look cheerful. We transferred the second mate of the Princess Charlotte and his boat's crew to the Emma. We captured four whales before we came to the land floe off Pond's Bay. All the other vessels made fast to the floe except ours and the Eclipse, of Peterhead, who stood off amongst the loose ice in the offing. Here we killed four more whales, and proceeded south to Coutts' Inlet. The whales being very scarce here, and not seeing any ships coming south, we returned north again. There we found the ships with as many fish as they could carry, some had six or seven alongside. The last run of whales had taken place twelve hours before our arrival. We made fast to the floe, and killed a large unicorn or narwhal, whose horn measured nine feet. I will now give one or two instances of what occurred while whaling here. We killed one, and had to tow it four miles to the ship. We had just sat down to take refreshment, when a fall was called, the captain had struck a whale, so had to leave our food untouched and go to his assistance. From the time she was struck until she was dead occupied twenty-six hours. She had taken us five miles from the ship. If ever there was a diabolical whale this was the worst. It would allow us to approach quite near, and when in the act of delivering a lance, she would strike out with her tail and fins in a most vicious manner. Although we had four harpoons in her, she did not lose her strength, and during the whole time it was perfectly calm, and the water smooth. Many were the hair-breadth escapes we had during the time. The mate's boat's crew and mine did not taste any food for thirty hours. Two other vessels in the neighbourhood fastened about the same time and encountered similar difficulties, one being worse off than we were. A bight of the line flew over a harpooner's head and cut him nearly in two. I think he belonged to the Pacific, of Aberdeen. To shew the rapidity of the whale, I struck one, and she immediately rushed under the floe down to the bottom. With the exception of a few fathoms, the lines were all run out in three and a half minutes. Each line is 120 fathoms long, and there are five in each boat, making in all 600 fathoms, or 3,600 feet, shewing her speed to be 1,000 feet per minute. She was hauled up dead, having broken her neck. The head had been embedded eight feet in the dark blue mud.

We plied to and fro, and gradually came south, sometimes at the middle ice, then inshore, and were often in chase of whales, but could not get fast to any. We sent our boats inshore on the south side of Home Bay, and saw several, then bore up for Cape Hooper.

A very large iceberg once lay off here and remained three years. It was four and a half miles in circumference, and was called by the sailors Ross's Berg. Captain Ross, in his first voyage, mentions this berg, which was found to be 4,169 yards long, 3689 yards broad, and 51 feet high above the level of the sea. It was aground in 61 fathoms, and its weight was estimated by an officer of the Alexander at 1,292,397,673 tons. On ascending the flat top of this iceberg it was found occupied by a huge white bear, who, justly deeming discretion the better part of valour, sprang into the sea before he could be fired at.

Dr. Hayes measured an iceberg to the north of Melville Bay. The square wall which faced towards his base of measurement was 315 feet high, and a fraction over three quarters of a mile long. Being almost square-sided above the sea, the same shape must have extended beneath it ; and since, by measurements made two days before, Hayes had discovered that fresh water ice floating in salt water has above the surface to below it the proportion of one to seven, this crystallised mountain must have gone aground in a depth of nearly half a mile. A rude estimate of its size, made on the spot, gave in cubical contents about 27,000 millions of feet, and in weight something like 2,000 millions of tons.

The vast dimensions of the icebergs appear less astonishing when we consider that many of the glaciers or ice-rivers from which they are dislodged are equal in size or volume to the largest streams of continental Europe.

Thus one of the eight glaciers existing in the district of Omenak, in Greenland, is no less than an English mile broad, and forms an ice-wall rising 169 feet above the sea. Further to the north of Melville Bay and Whale Sound are the seat of vast ice-rivers. Here Tyndall Glacier forms a coast line of ice over two miles long, almost burying its face in the sea, and carrying the eye along a broad and winding valley, up steps of ice in giant height, until at length the slope loses itself in the unknown ice desert beyond. But grandest of all is the magnificent Humboldt Glacier, which forms a solid glassy wall 300 feet above the water level, with an unknown depth below it, while its curved face extends fully sixty miles in length. In the temperate zone it would be one of the mightiest rivers of the earth ; here in the frozen solitudes of the north it slowly drops its vast fragments into the waters, making the solitudes around re-echo with their fall.

The wonderful beauty of these crystal cliffs never appears to greater advantage than when clothed by the midnight sun with all the splendid colours of twilight glittering in the blaze of the brilliant heavens, seeming in the distance like masses of burnished metal or solid flame.

In the shadow of the bergs the water is of a rich green, and a deep cavern near exhibits the rich colour of malachite mingled with the transparency of the emerald, sometimes a broad streak of cobalt blue runs diagonally through them.

In the night the icebergs are readily distinguished even at a distance by their natural effulgence, and in foggy weather by a peculiar blackness in the atmosphere.

We shot several bears, and brought two on board alive. One of them was not quite half grown. After securing it to a ring bolt in the deck, and in the act of taking the other on board, a man accidentally let go the rope which held the former, and the deck was cleared in a very short time. Some of the sailors took refuge in the rigging, others below, and it was really laughable to see how quickly Bruin had the deck to itself. Though only a small animal, no one cared for a bite. It was immediately lassoed and secured in a cask for transportation home. A fine bear would fetch about £35 for a zoological garden.

Arriving in Cape Hooper harbour, we followed the general routine of rock nosing, most of the boats going to the outside, the others watering and preparing for the passage home. We were joined by the Diana, of Hull, which had got full of seals at Greenland, and been home to discharge, then came here whaling. Very few fish were seen, it being October. The weather became very unsettled, and the young ice was rapidly making in the smooth sheltered inlets. We took advantage of the full moon to light us out of this cold inhospitable country, and got under weigh in company with other ships which had also made a good voyage, and were homeward bound. We arrived at Lerwick after a favourable passage, received letters from home, discharged the Shetland men, and in due time arrived safely at Hull.

CHAPTER X.

IN 1857 I went as mate of the barque Emma, with
Captain Parker. This vessel was built of teak wood,
in the East Indies, and was bought by Thos. Ward, Esq.,
of Hull, who had her doubly fortified for the Greenland
whaling trade. She was a fine weatherly craft, and as handy
as a cutter. No vessel could be better fitted out. We had
eighteen months' provisions on board, with a good supply of
best canister meat from Morton's, of Leith (one of the few
firms which then provided such food). We had pickles and
many dainties which other ships had not. She was fitted
with iron tanks to hold the blubber, which is a great
improvement from the old system of using casks for that
purpose. We left Hull for Stromness, in the Orkney
Islands, to obtain our complement of men. We brought up
in the harbour, which is a small one. Vessels usually lie
at the back of the Holms, i.e., a sandbank which lies
parallel with the town, but in bad weather a heavy swell
rolls in, which obliges the ships to ride with both anchors
down. It is a long way from the town, which is very
inconvenient. When the weather is fine, and the water
smooth, it is so clear that the anchor and cable can be
plainly discerned at the bottom, also shoals of fish called
sillocks, which resemble a large sprat, but more delicious.
The northern islands abound in them so plentifully that some-

times they use them for manure, after extracting the oil from the liver. The second year they are called pelticks, and are the size of a haddock. The islanders say after that they seek deeper water, and become full-grown coalfish. Whatever they may be, they are exceedingly sweet. Now that regular steamers run direct from Leith, I have no doubt during the fore part of the year they would find a ready market in this country.

When the ship was ready for sailing, the Union Jack was hoisted, the topsails mastheaded, the bellman went round the town announcing that the ship was ready for sea. In the short space of one or one-and-a-half hours the men were all on board with their chests, beds, and clothes. The anchor was weighed, watches chosen, and again we were on our way to seek a living surrounded by so many obstacles and difficulties. There cannot be a more speculative class of seamen than the officers who follow this occupation. If nothing is captured, they come home from £3 to £5 in debt to the master for small stores, etc., and those who have families to maintain, instead of enjoying their homes for a short time, have to join ships in the coal trade for the winter. A bad voyage also entailed serious loss to the owners. Having had a fair run out, we fell in with the ice in Lat. 62½° N., Long. 56½ W., at the old S.W. fishing ground, Resolution Island being about 240 miles distant. The ice, although very heavy, lay in streams from the pack, and was favourable for whaling, but none were seen. After cruising about two or three days we worked our way to the northward, through streams of ice which gradually led us towards the land, to a place called Goodhaal, at the entrance of Baals River. From there we ran along the land among straggling heavy ice, with clear weather. On reaching Holsteinberg the natives came off and told us that westerly winds had prevailed for a long time. It was therefore expected we should have easterly winds in the

summer, when we most needed them.   Our usual haunts
were visited, but only two whales were captured.   One had
a gun harpoon embedded in its body, which must have been
there many years, as the name of the ship to which it
belonged could not be deciphered.   The part which pro-
jected from the body was worn to a point, and all round it
the blubber was quite hard, and the stench was so bad that
we threw the harpoon overboard.   The time was approaching for us to seek a passage through Melville Bay.   When
off Proven we fell in with the Gipsy and Undaunted,
both belonging Peterhead.   Our captain being the senior,
the others came on board to hold a consultation, so it was
decided to keep company with each other through the Bay.
A strong S.W. wind began to blow.   We led the way, and
had just passed Upernavik.   In the vicinity of the great
glacier many sunken rocks lay in our way.   A good look-
out was kept on the fore topsail yard, and the master
in the crow's nest.   Yet with all his experience and precaution we ran upon a rock which was covered with a piece
of ice.   At the time we struck we were going at the rate of
eight knots.   All hands were called to lighten the ship
forward.   There was five fathoms water aft, and four on the
starboard beam.   The fore part and port beam was fast on
the rock.   The captains of the other vessels kindly offered
their assistance, but our master told them they had better
leave us and make the best of their way northward while
there was water.   They returned to their respective ships,
but soon made fast to an iceberg five miles away.   Both
being strangers to Davis's Straits, they did not like to
venture any further, seeing it was such an intricate part, and
one of them had already glided over a rock.   There is very
little rise and fall in this part, and what is called a tide and
a half tide.   The day tide being the best, we took warps
out to a berg astern, and provisions from the fore hold, and
the following day hove her off.   Although the crew had

been on duty such a long time, and were thoroughly tired, we once more set our canvas and ran north until we came to an island called Kinatuk, where the ice was tightly packed. We made fast to a berg close to the island, and were joined by our companions. I was sent on shore to look at the grave of an Esquimaux whom Captain Parker brought to England with her husband from Cumberland Gulf in 1847. She died on the return journey, and was buried on this island. The grave was undisturbed, and the head board, which had her name, etc., inscribed upon it, was in good preservation. She was called Ukaluk, in English, a hare. Her husband's name was Memeadluk. He still resided in Cumberland Gulf, leading a lazy life. He was too idle to hunt or fish any more so long as the presents lasted which he had received in England. We lay here two days, when the ice opened, and with towing we came to some open water. A light favourable breeze springing up, we set our sails and came to Horse Head, a peculiar headland or island which lies a little north of Cape Shackleton, in Lat. $73\frac{1}{2}°$ N. In due time we came to the Duck Islands, where we met with a block. The S.W. wind had jammed the ice tight upon a reef of bergs which lay a little north of the islands. As it became calm we commenced to tow, and were in such close proximity to a very large rugged, splintered berg, that we had to brace our yards sharp up to pass it, and the men refrained from singing for fear the sound might cause the berg to split and fall upon us. I have seen one split in two from the stroke of an ice drill, which caused the death of three men. The report of a gun will sometimes have the same effect, as they are very brittle, especially when aground. We reached Melville Bay with the Devil's Thumb well open off Wilcox Point. This peculiar mountain lies a little inland, and has the appearance of a hand stretched out with the thumb upright. The ice being close, we made fast to the land floe with our companions the Gipsy and Undaunted, and were in good hopes of getting through the bay in a reasonable time. Anticipating easterly winds and calms, we got provisions and two splendid tents on deck in case of emergency. This was the first time such comforts had been provided for the crew, and were made expressly for the purpose. One would accommodate forty men, and the other twenty. The poles were in six feet lengths, and fitted with copper sockets. During the time we were in the bay we occasionally used them, and they were highly appreciated. The next calm opened the floes, but we only got two miles further when the ice again closed. A dark nimbus sky rising in the south, we began to saw a dock, but had scarcely time to get into it before a heavy gale burst upon us, which made us fear we should lose our good ship. The ice had not much space to drive, otherwise the floe would have crushed us to pieces. After this came a succession of gales for six weeks from the same quarter, quite contrary to what we anticipated. At times when the weather was very clear we could see the southern fleet from the mast head, sheltered under the lee of the Duck Islands. Every possible means were used to secure a safe place by sawing docks. During the time we were in the bay we sawed eighteen, but each gale broke the floes more and more, and left no safe place for us. One heavy gale caused the floe to pass over the Gipsy and Undaunted, the ice cutting through them, and they became total wrecks.

We were forced on our beam ends, and every moment expected to share the same fate. However, we providentially escaped, and prayed to be led safely out of the much dreaded Melville Bay. Our critical position, and the loss of the other two ships, made some of our men very depressed. The crew of the Undaunted set her on fire, which is the usual practice when a whaler becomes a wreck, and the crews of both ships came on board of us and stayed two days to prepare themselves for launching their boats in order to reach the southern fleet. We assisted them three or four miles on their journey, and when parting gave them three cheers. Three of the Gipsy's men preferred staying with us.

Being now left alone, our thoughts were naturally sad, but it does not do to look upon the dark side. We had a number of sledges with us. The runners were made of African oak, which does not splinter or tear, but becomes smoother with constant wear. To keep the men employed, each watch took one of the sledges and an empty cask to seek water from the neighbouring bergs. From some we got a few buckets, from others a larger quantity. Sometimes it was a good four hours' work to fill one cask. Bears were often seen, and several shot. An officer always accompanied each sledge with a rifle and a couple of lances. On one occasion a bear came boldly up within ten yards of us, when I fired and shot it through the head, fortunately killing it. Some of the men were already on their way to the ship, which lay five miles off. In such a case it is of no use attempting to run, because Bruin's legs will carry him much faster over the ice than ours will, therefore it is better to reserve your strength to meet the emergency. Another time I was sent along the floe edge at one particular place to see if there was any movement in the ice. The weather at the time was very foggy. Whilst walking I jumped off a hummock of ice alongside a full-grown bear, lying asleep. It was awake in a moment, evidently as much surprised as I was. There was neither time nor room to take a deliberate aim, but I fired, and the bullet entered its brain. I thought this adventure quite sufficient for one day, and retraced my steps to the ship, as there were so many bears' footprints in the snow, that it was not safe to be alone. I returned with some men to drag the carcase to the ship, after seeing that the ice was stationary.

One morning, as we were having breakfast in the cabin, after a severe gale and ice pressure, a shot was fired from the deck. We rushed up to ascertain the cause, and found the harpooner on watch had fired at a bear, but missed it. The captain ordered me to take my rifle and follow it. The animal walked leisurely away. It was half a mile from the ship before I caught up with it. When within ten yards, which I considered quite near enough upon the ice when a man has only one charge for his gun, and that a muzzle loader, the brute lay down with its head between its paws as though in the act of making a spring if I should come closer. I aimed at the eye (which is the best mark when end on), at that moment it moved its head, and in consequence the bullet entered the shoulder, breaking it, and making the blood spout out in a stream. I immediately made tracks for the ship to get more ammunition, but when half way I met a boy bringing me a lance. Giving him the empty rifle, I returned to the bear. When I came close to it the animal rose on its hind legs. I struck at its breast with the lance, but it was so blunt and weak that it bent double without entering the body, causing me to fall near the animal, which immediately grasped me with its teeth on the right thigh, biting a piece out of my trousers, and inserting its four tusks in my flesh. At the same time three or four bullets came whistling close to me, which were recklessly fired by some men the master had sent to assist me when he saw my position. However I scrambled out of Bruin's reach. All this took place quicker than can be related. The bear now began to retreat, but could not make much progress on account of the wound and loss of blood. By this time the men had come up. I took one of the rifles and shot the bear through the head. The bite did not incapacitate me from duty, but it was three or four weeks before the wound was completely healed. I still carry the marks. When I returned on board, the master found fault because I had made a hole in the skin. He said I ought to have shot it through the brain at first. Evidently he did not consider my skin. He was an excellent marksman himself, and expected me to be the same, yet under the circumstances I considered I had done very well. However, we who were brought up to Arctic life at the period of which I am writing, gain experience as we grow older not to act foolhardily. At the same time I think youths ought not to be checked in showing their pluck in times of danger. It is better to let them gain self-confidence and self-reliance.

For two weeks longer we were tightly jammed up without any prospect of being liberated, and some of the men began to get disconsolate. Our boatswain's work was all done. The old rope was made into gaskets, foxes, spunyarn, etc. Our three suits of sails were middle stiched, rigging in good order, and there was very little more to occupy the men. We therefore renewed our travels with the sledges in order to drive dull care away. I must confess that the times were not lively, but our position could not be avoided, and we dare not use our ice saws for fear of weakening the point of ice which sheltered us. The southern ships had all disappeared save one, which we hoped was the Isabel, a small screw steamer, brigantine rigged, which had formerly been on discovery, and had now come out to be a tender to tow us in calm weather. It was Captain Parker's intention, if not successful north, to seek for whales in the neighbourhood of York Bay or Frozen Straits, in Fox Channel. For three days we had calm weather ; the sun shone brilliantly, yet we could not shake off the depression of spirits until a crack in the ice made its appearance. Then all was bustle and excitement getting provisions, etc., on board, and preparing for the first slack to liberate us. Our ship was ice bound, and it took some time to get the ice off the ship's bottom, which we did by taking a line from the topmast head, and part of the crew took a run upon the ice to make her roll backward and forward until the ship was clear, when we began to track to the southward. The following day we reached the Duck Islands in company with the Isabel, from which we received letters, papers, and little delicacies which our friends had sent us. Our Melville Bay troubles were soon forgotten. With our sails set and in clear water our spirits rose like a barometer after a gale, and we made the best of our way further south, judging there would be no opening in the ice until we got down to Cape Searle or thereabouts. Such was the case, and we crossed over to the West Land, arriving shortly after the other ships, which left some time before us. We got into the west water off Cape Searle, and proceeded to the northward. Fogs in the Arctic Regions are very frequent, and sometimes last more than a week. A good prospect may be before us when a dense fog suddenly sets in, and we are liable to take a wrong opening, thereby losing the only chance of a successful voyage. *Nil desperandum* was our motto. We sailed on until we arrived off Scott's Inlet, and dodged to and fro, but saw very few whales. Two or three were fired at, and missed. We cruised about with the Isabel among the middle ice, the other ships having gone further north towards Pond's Bay, and, if possible, to Lancaster Sound, but the season was too far advanced for that place. One day we saw a great number of sword-fish. Whenever these appear, no whales, unicorns, or any other denizens of the deep are to be found. The northern sword-fish is one of the largest kind of the whale family. I have seen them about 25 feet long, or perhaps longer. When those fish are seen on the fishing grounds, whales, seals, etc., vacate the place. They have a long, sharp, upright fin on the top of the back. Those fins are about six or seven feet long on the larger fish, which has earned for them the name of sword-fish. These enormous creatures come to the surface to breathe, like other whales, and move more in a zig-zag

direction along the top of the water, so that they may
be seen from three to half-a-dozen times, though only the
upper part of their shining black bodies.   First the snout is
to be seen, and from the " blow-hole " is forced the breath ;
then the black fin, or sword, and the hinder parts of the
back make their appearance.   The tail is only seen when
the creature dives under.   The Dutch name it the
" Nordkaper."   The Americans call it the " Killer."   The
Norwegians " Whale-catcher," and the English call it the
" Sword-fish."   It is confirmed by the least prejudiced
whalers of the present day, that they are the whale's
greatest enemies, following and attacking it in droves.   The
sword-fish in the northern latitudes are unlike those in the
south. The former have a large triangular fin on the middle
of the back, with which it torments the whale from below.
Their appearance made our captain determine to proceed
at once to Cumberland Gulf.   On arriving there we went
up the north side, and were taken in tow by the Isabel up
a fiord, called by the natives Panatung.   It was not quite
dark when we entered it; a bright moon was shining, and
we hoped to get an anchorage close to the outside.

The Fiord was not more than half a mile wide, and the
land was very high on each side.   All hands were on deck,
but soundings could not be obtained at sixty fathoms.   This
was the first time a ship had been in this place.   The
Isabel continued to tow us at the rate of about one mile per
hour.   There were no signs of being able to bring up.   At
daybreak, we saw a low sandy point, which we supposed to
be the end of the Fiord.   However, we steamed round it,
and brought up about seventeen miles from the outside.
The next day I was ordered to take my boat, and try to find
how far it extended, as it appeared to run many miles
inland.   Having provisioned the boat for four days, we
started at three a.m., and set our sail with a nice breeze up
the narrow channel.   We passed several points of land, and

halted at an ancient Esquimaux summer resort. The grass
had grown over the stones which had held their sealskin
topecks down. The whole place was strewed with the skulls
of seals, walrus, deer, and foxes. It must have been a
glorious feeding-place for the natives. They always pitch
their tents near a running stream of water. This was the
first time in my life that I felt the bite of a mosquito;
indeed, I should not have known what insect it was, if a
sailor, who had been in southern latitudes, had not told me.
It was as strange to me as thunder or lightning, which I had
not heard or seen during the time I had been at sea. The
Fiord appeared to terminate a few miles further on. It was
getting late in the afternoon when we got within a mile of
the end. Our boat grounded in the middle, which was only
a quarter of a mile wide, and in a short time we were left
high and dry and a long way from the water's edge. The
rise and fall was very great, and the place very shoal.
Therefore we travelled over the sand to the shore. The
scenery was exceedingly grand. At the head of the place
was a beautiful cascade with three falls, each one having a
descent of about thirty feet. When walking toward it, we saw
forty or fifty deer, but they were too far off for a shot. They
had already seen us, and it was amusing to see them bound-
ing over the rocks and then stand for a few minutes, looking
at us as if wondering why we were there. Being very
thirsty, we lay down to take a drink from a stream of water.
My boatsteerer called my attention, and said "There is gold
here." On examining it closely, we found the black sand
at the bottom strewed with dust which looked very much
like gold. He had been at the Californian gold diggings. I
took it for granted such might be the case. It is still my
opinion that the country is wealthy in minerals of all kinds,
but perhaps not to such an extent as to allow it to be worked
without a great expenditure of capital. We are not expected
to be judges of metals. The man took a mitten full on

G

board, and gave it to the captain of the Isabel, but I heard
nothing more of it.   There is a great deal of ironstone and
mica here.   The granite rocks are studded with garnets, but
their long exposure to the frost makes them worthless.   I
have often chipped them out as large as a pea.   On taking
a view from the top of a hill, we saw a level plain with deer
grazing, but at too great a distance for us to chase them
before darkness would overtake us.   As the tide was beginning to flow, we walked quickly back to the boat.   On our
way we gathered some pretty little flowers.   They were
small, but would look well in groups in a garden.   I did not
know the name of them.   They were lovely, and we
treasured them, as they were a novelty to us.   We arrived
back in time to get our coffee before the boat floated.   It was
a clear moonlight night, so we pulled down the Fiord, and
arrived back to the ship at four a.m., having been away
twenty-five hours.   I drew a map of the entire Fiord.   We
remained here two days, as it was too far to send the boats
away to the outside.   Steam was got up by the Isabel,
which took us in tow.   On the north side we found anchorage in fifteen fathoms, which we had not seen when going
up in the dark, and only three miles from the outside.   The
boats went away as usual for three or four days, but no
whales were seen.   I was sent to look for another harbour
along the coast, and to draw an outline of the land and take
note where rocks lay.   I went on shore at various places to look
for them.   The water was so clear in places that rocks could
be discerned at a great depth.   No suitable harbour was found
near.   Steam was got up, and both vessels proceeded to the
opposite side of the Gulf to Niatlik.   One or two fish were
seen, but they were going at a rapid rate to the southward.
Here we were joined by the Traveller, belonging to Peterhead, and the brig Clara.   They had a small screw steamer
called the Jackal, and were apparently on the same errand
as ourselves, and taking the same route.   The master of the

Traveller and ours were friendly, but each kept his own counsel. It was most singular, whenever we found a new harbour they were sure to come directly after. We were in twelve different harbours where no ship had ever been before. The land extending from Cumberland Gulf to Frobisher Straits had not been then explored, but was called Meta Incognita. At the N.W. part of the Gulf lies a large lake called Lake Kennedy, but by the natives Nitiling. I never heard of any civilised person being there. The natives gave me a sketch of it, and said there were three large waterfalls which emptied themselves into the Gulf. It is marked down on our charts, but, no doubt, they have been furnished by the Esquimaux, as the one I have from them corresponds with the others, excepting the waterfalls. We gave names to many islands, capes, and bays, and then came to a place called Nugumut, where the natives are taller and more wild than those at Kemisuack. Some of the latter told me that formerly the Nugumut natives made raids upon them, killing the men and taking the women away. I cannot vouch for the truth of this statement, yet I noticed they were very distant with each other when they met. If both tribes happened to encamp for a short time in the same place, they always left a large space between their huts.

We were now always among natives ; some had never seen a white man or a ship before, although they had heard of them from those who had been to Hudson's Bay. Many of the men had a moustache, and their hair knotted or tied on the fore part of the head. Their kamicks, or boots, were also different about the ankle, and the women were more tattooed. The Nugumut (innuit), or natives, are great thieves, but the poor things are much to be pitied. Their constant cry was "peletay," which means give something. Although they would not eat our food, they would beg it and leave it behind a spar, or elsewhere. Their canoes were larger than those belonging to the natives further north. Some would carry three persons. This tribe are like the others, of a roving disposition, and have no permanent abode until the winter sets in. Upon the whole they are cleaner than those at Cape York and Pond's Bay. Our last anchorage was near to a narrow fiord, which led into Frobisher Straits. One day I was sent to take a survey of it. When half-way through the fiord I found part of the hill side composed of stones of all sizes, resembling Turkey stones, which are used for sharpening tools. I brought a quantity of the smaller sizes on board, and it was most remarkable that in such a remote region they should have the appearance of having passed through the hands of a mason. All the other rocks were granite and ironstone. The fiord was eight miles in length, and opened midway in Frobisher Straits. I travelled up the mountain, and had a good view of the greater part of the straits. It was very rocky, and so far as I could judge it was about thirty miles across. There were no animals in sight, or any signs of it being visited by natives. On my return I reported to the captain that the fiord opened into the straits as he anticipated. If our business had called us in that direction it would have saved us many miles. The winter was fast approaching, yet with our little steam companion we remained longer than we otherwise could have done. As there was no prospect of getting any more whales, we weighed anchor and sailed for home. Therefore the captain's intention of going to Fox Channel was abandoned for the present. We much regretted that we had not gone direct there when we came from the north. Fox's Channel was so called by Captain Luke Fox, or North-West Fox, once a Younger Brother of the Trinity House, Hull, who sailed from the Thames in the year 1631, the same day as Captain James sailed from Bristol, both in search of a north-west passage. We eventually arrived at Stromness, and stayed there two days for a favourable wind, which brought us safely to Hull.

CHAPTER XI.

IN 1858 I again joined the Emma as mate. The fitting out and the passage across the Atlantic was similar to the previous voyage until we entered the Straits. One evening, after a fresh gale, the wind suddenly lulled, and the sails flapped to the masts. The Aurora Borealis burst out in magnificent array, enveloping the ship in light vapour of various colours. It lighted up the vessel so that the smallest print was quite discernible. This display continued two hours, after which a heavy swell began to rise from the S.W. The barometer fell rapidly. We furled the topgallant sails, and put the ship under close reefed topsails, and preparations were made for a coming gale, which was soon upon us, and ran us as far as the Arctic Circle. Then the gale ceased as quickly as it came, leaving a very heavy swell. We were surrounded with innumerable icebergs. Our sails were frozen stiff, and we now required all canvas set to keep us clear of them, also the heavy pieces of ice, on to which the swell was drifting us. All hands were called to set sail, which took us a long time, on account of their frozen condition. We were in a most serious predicament, the seas reaching half-way up the bergs. Fortunately a fine breeze sprang up, which enabled us to thread our way among them and take an inshore passage. I believe it was this gale which liberated the Fox, Captain McClintock,

after his long dreary drift of six months down the straits.
Arriving off Disco, and having despatches from Denmark,
and luxuries of various kinds from England for Major
Olrick, at Goodhavn, I was sent with two boats to deliver
them to him, but finding the ice blocked the harbour, I was
obliged to land at the point and travel overland to the
settlement.   It was 3 a.m. when I reached the Governor's
house, but everybody was stirring.   I may here remark that
at this time of the year it is never dark.   It gave me great
pleasure to see how delighted they were to receive letters
from their absent friends in Denmark, having been so long
cooped up in such a solitary place without any news.   This
settlement is situated on a long low projecting point of
land, at the end of which is a look-out, or signal house.   At
the back the land is very high and precipitous, so that at a
short distance off the settlement cannot be seen.   It is
sheltered from all winds.   The natives at this place will not
travel up the high land.   They say that his Satanic Majesty
roams about the hills, with other evil spirits.   The entrance
to the harbour is from the north, through a narrow channel,
with a rock in the middle, and ships generally tow in, but
very few of the whalers put in there.   The place is
exceedingly picturesque.   Its beautiful harbour, with the
neat black wooden houses of the resident Danes and the
Esquimaux huts, shows to great advantage.   About ten days
afterwards we were in pursuit of two whales, which led us
close to the settlement.

The ice had now cleared away excepting a small piece
that remained attached to the land.   Our boats were placed
round the harbour, making a very pretty scene.   All the
people belonging to the settlement were anxiously watching
us.   They said a whale had not been seen in the harbour
for upwards of twenty years.   I expect the whales saw us on
account of the water being so shallow, for they moved from
one part to another, and only gave one blast when they rose

to the surface to take breath. However watchful we were, they contrived to escape from us to the outside. I distinctly saw one pass under my boat, which lay at the entrance. She was swimming on her side, evidently watching our movements, but at too great a depth for the harpoon to reach. We cruised off and on for several days longer, and saw many whales, but could not get near them. One day as the ship was standing towards the land we fell in with four boats belonging to the settlement. All of them were entangled to a fish, which was making for the offing. Undoubtedly they would have lost it, as the breeze was freshening, and a cross sea was rising, but the captain, with his usual kindness in such cases, ordered me to lower a boat and lance it for them, which I did, and soon killed it. Their boats took it in tow towards Goodhavn, but the wind was contrary, so they towed it into a small cove, waiting for the wind to change, but by the time they reached Goodhavn their prize was nearly eaten by sharks. I think it would have been better on board our ship. The sharks would then have missed a good treat. Perhaps I may be permitted to remark that there are several places on the east side where it would be most profitable if an enterprising English company were allowed by the Danish Government to prosecute the whale fishery in the early part of the year, and were provided with steam launches to follow the boats and take them in tow through the patches of bay ice, and also to tow the whales when killed. One place in particular lies between Fortune Bay and Goodhavn, in which the Emma wintered in 1855-1856.

On the east side fishing grounds we were much disappointed because we saw so many whales but could not secure one, although our boats were constantly in chase. Nothing of importance occurred until we reached Upernavik, where the ice was tightly packed, not only upon the outlying islands, but as far north as the eye could reach. We made

fast to an iceberg close to the settlement, and shortly afterwards the Fox, Captain McClintock, came alongside.
He remained with us three days. We provided him with
ten tons of coal, some beef, blocks, cordage, wheelbarrows,
and baskets, which we had on board to spare. The latter
articles with which we supplied him were taken out by the
Emma in 1855, when they intended to winter in Niatlik, for
the purpose of laying a broad track of gravel on the ice
from the ship towards the outside, to weaken it and to keep
the men in exercise ; for instance, if a rope yarn be laid on
the ice twenty-four hours its weight will sink it half-an-inch.
During the last two days the weather was calm, and the ice
began to open in all directions. Away steamed the Fox
on her risky mission in search of Sir John Franklin. All
was now bustle and activity—the ships casting off from their
respective bergs and towing with their boats. The little
Fox soon steamed out of sight. While we were detained
here I was generally away shooting birds or anything eatable
for a treat. The eider duck and loom are the principal
birds. The latter are more numerous. Their flesh is very
sweet, especially when made into sea pies. This bird
differs from those caught at Flambro' Head, which have a
very fishy taste. Those in Davis's Straits are free from such
an objectionable flavour. They are not plucked like other
birds, but skinned, which is soon done. There is another
pretty little bird called the dovekie, about the size of a
young pigeon. It is jet black, with the exception of a small
patch of white on each wing and bright red legs. They are
so quick that they will dive before the shot reaches them.
The auk is generally called "loom," and Sanderson's Hope
is the largest loomery on the east side of Davis's Straits.
There is also the little auk, called "roaches," exactly like the
loom, but it is only the size of a sparrow, and their breeding-
place is at the Crimson Cliffs, near to Cape York. They
fly in large flocks. I have brought down thirty-five with

one charge of sparrow shot. The natives at that place catch them in a net attached to a stick as they fly past the rocks. When ships are detained on the east side until July they often send their boats among the islands at the foot of the glaciers, and I have known them gather one hundred and twenty dozen loom and eider duck eggs at one venture. Such times, however, are not profitable, but have a sorry look-out for the whaler when pay-day comes. With the ice opening, the ships made good progress in towing. The weather was calm and clear until we reached the north part of the group of Vrow Islands ; then a fine, easterly wind sprang up. We set all sails, and took the boats on board after towing sixteen hours, and were very thankful for a rest, being very tired. We sailed, towed, and tracked, and eventually arrived in Melville Bay. In due time we got into the north water off Cape York, working up to Cape Dudley Digges before we rounded the north point of the ice, and then crossed over in sight of Coburg Island at the entrance of Jones' Sound. Everybody was in high spirits with the expectation of getting a full ship. Harpoons were cleaned, lances and knives sharpened, guns prepared, and everything in readiness for whaling. A fine breeze blowing from the northward, we sailed merrily towards Lancaster Sound, when suddenly we came to a solid body of unbroken ice, which formed into a large, deep bight. We sailed to the end of it, and then worked north to find a passage to the westward, but there was no outlet. East, west, and south presented in turn nothing but large, unbroken fields of ice, which sight damped our spirits. A consultation was held by the masters, which led to the conclusion that it was better to return through the bay, and seek a passage further south. The Jane, belonging to Bo'ness, and the Heroine, of Dundee, were lost in Melville Bay this year, but we and the others got safely back through it with very little trouble. When off Duck Island, we encountered a gale of wind from the S.W., with thick snow and a nasty sea, which a short time previously had been covered with ice. It is almost incredible what great changes will suddenly take place in this country.

To the westward of the islands a deep bight remained, in which I afterwards heard that Captain McClintock entered after he had left us at Upernavik, and arrived easily on the west side. If we had known that, our prospects would have been different, but in sailing ships our safety lies in having hold of the fast ice or land floe. Getting beset amongst loose floes or in a pack is sometimes a serious thing. However, we kept working our way south until abreast of Cape Searle. The weather had been so foggy that we could not get sight of any slack places in the ice until now. We succeeded in getting into the west water, and proceeded north until we came to Agnes Monument, a little north of the river Clyde, which is a noted place for whales. Several were seen, and two were fired at without effect. We cruised among the middle ice, which lay in a favourable position. Only a few straggling fish were seen here. Some ships got one or two. The weather continued so foggy that we could do nothing but ply to and fro. We shot several bears. One particular event occurred which shews that such savage animals have parental feelings. I went after an old bear and two large cubs, which were in open water. The young ones could not swim fast enough, so the old bear always kept between us and her cubs, occasionally turning round to growl and give the young ones time to get further away. So long as they were swimming in the direction of the ship we made no attempt to kill them, as it would save us the trouble of towing. Suddenly the old one turned upon us, and I had scarcely time to shoot her before she was at the bow of the boat. The young cubs, seeing the mother lie motionless, turned back and dived under her head to lift it up, but, finding she would not respond to their caresses, they boldly charged us. I shot one, and killed the other with a lance, as I had not time to reload.

A few days later our boats were sent ashore on the south side of Home Bay in search of whales. I was pulling through a stream of ice, and saw two large bears asleep upon a heavy piece, evidently having enjoyed a good meal off a bladder-nose seal. One woke up, and plunged into the water. We chased it, and the second mate, seeing us pulling, followed us. When he saw the other bear lying on the ice, he thought I had shot it, and prepared to take it on board ; two men were going to put a rope round its neck, saying, " Get up, Jack," which order he at once obeyed, to the dismay of the whole party ; but, fortunately, the second mate had his rifle at hand and shot it. I also killed the other one. The ship gradually drifted further south, until we reached Cape Searle, when six boats were sent inshore in quest of whales, and to gain information from the natives whether there were any in that neighbourhood. We pulled into a snug harbour, about seven miles from the ship, called Hangsman's Cove, so named from a native being found hung by the neck from a perpendicular rock. Whether he had committed suicide, or had been punished by his own people, was never known. On landing where the natives had encamped, a shocking sight presented itself to us. There were several summer huts standing, but the occupants were all dead ; but how they had come by their death was a mystery.

Some lay in their huts, others outside, numbering twenty altogether. They were partly eaten, either by wild animals or their own poor, starved dogs. We saw something high up the hills, but from the distance could not tell whether they were wolves or dogs. We did not stay long on shore, but pulled towards the ship, and killed a walrus on our way, and flensed it on a piece of ice. We did not see any whales, so the following day sent our boats to a place called

Durban, and continued to coast along, each day sending
boats inshore until we reached Cape Dyer ; then sailed for
Cumberland Gulf, and brought up at Niatlik, but not seeing
many whales we got under weigh and proceeded to Nugumut,
and came to anchor in one of the harbours on the north
side, having taken some natives as guides.   Although they
know the land along the coast for many miles, they have not
the least idea of sunken rocks or depth of water.   This time
we did not go outside, but threaded our way among the
islands, thereby saving much time and anxiety.   We had no
previous knowledge of this route.   In this harbour we found
a very large encampment of Nugumut Esquimaux.   The next
day when we went to the outside we found a heavy swell rolling
in, and breaking so high upon the mainland that it was impossible to go near.   The whole coast was studded with rocks,
some extending more than a mile.   The wind was off the
land when we came into this harbour, and we soon found
out that it was not a likely place for whales.   The master
gave orders if the weather was rough to return on board.
One morning the boats went out, but were forced to come
back at 10 a.m. on account of a strong easterly wind and
heavy sea.   I was surprised to see the native omeacks or
luggage boats (so called because they carry the tents, women,
and children from place to place) alongside with bows and
arrows fixed in the gunwales of their boats.   There were no
women or children to be seen, which we thought strange.
When our boats were hoisted up one of the crew told me
that some of the natives had their knives or bone-handled
daggers inside their jackets.

When I mentioned this to the master, he ridiculed the
idea of the natives attempting to do us any harm.   One of
our men took hold of a native in a playful manner, when he
immediately drew his knife.   Then I was ordered to send
them on shore, which was our usual practice when night
came on.   I hurried them off, and thought no more about

the affair. The following morning was calm, and the boats were sent away again. It was my turn to stay by the ship to fill up our fresh water tanks from the land. Having pulled to the beach, I was much surprised at the natives refusing to allow me to land. Their attitude and shouts were apparently hostile. They seemed inclined to retaliate for sending them on shore the previous day. Therefore, I had no alternative but to return. They did not come on board until late in the afternoon, when they appeared as friendly as ever. I was again sent on shore with a cask for water, and this time they gave us a helping hand to pass the buckets along, which was very unusual, as they do not like too much work. This was the first and only time I saw any unfriendly feeling manifested by the Esquimaux. What their first intentions had been we never could ascertain. I once read of them taking possession of a ship belonging to the Hudson Bay Company, but I never thought of their daring to attempt it with us, who numbered fifty-two men, though when seven boats are away, it leaves but few in charge of the ship. The next day all hands were told off to fill up our fresh water tanks, as it was freezing keenly. The captain sent me on shore with two of the most ill-favoured looking natives amongst the lot. They told the captain that a short distance inland were plenty of deer. We travelled four hours, then came to a small lake. Footprints were numerous and quite fresh, but no deer were in sight. We remained two hours longer ; then I wished to return on board before it became dark, and told them my intentions, but they flatly refused, saying we must stay until dusk, when the deer would come from the hills to drink. I was rather puzzled how to act. If it took us four hours to come, and we had shot anything, it would have taken five hours at the least to return with a load. The captain had lent each native a gun, which was the first they had seen, although they had heard of them.

I kept the ammunition, and tried to persuade them to
fire their guns at a mark, but they said the deer would be
alarmed, and would not come near us, which was quite
correct.   I did not feel safe in their company so far from
the ship.   At last I promised if they would return at once I
would give them a knife.   They readily accepted the offer,
and I was glad to be able to come to such terms with them.
We returned to the ship by a route which only took us two
and a half hours.   It was dark when we came to the beach.
When I got on board I gave them the knives I promised,
hoping never to be accompanied again on the land by them.
As we did not see any whales in this place we got under
weigh with a nice breeze, and before dark anchored in
another harbour about ten miles away.   In these waters it
is impossible for a ship to keep under way during the night.
The rocks are so numerous, and the rise and fall of the tide
so great, we must decide beforehand where the next harbour
is situated.   We possessed a rough chart, sketched by a
native woman, which is in my possession at the present
time, with a few improvements I have made.   We also took
two or three natives with us as guides.   This place had been
a favourite resort for them at some remote period, judging
from the remains of numerous summer and winter huts,
also a large quantity of walrus bones and skulls lying about,
from which the tusks had been extracted.   I have often
wondered from where the natives in this neighbourhood
obtained their iron.   Most of it appeared to have been iron
hooping converted into lance heads and harpoon points, but
that used for knives was four inches broad.   Some of them
said it had come a long distance from the south ; others
that they found it on the beach, which would probably
have been taken from casks or wreckage that had drifted
ashore.   When a native dies they invariably bury all he
possesses with him.   I think with such a scarce article as
iron this may not be done, but that it is handed down from

one to another. I have also been astonished why the natives do not leave the walls of their winter huts standing. Is it selfishness, so that other natives may not occupy them, or may it be superstition? We had no sooner brought up, than two luggage boats and ten canoes came into the harbour, but they could not be persuaded by us to come alongside. The natives who were on board spoke to them, and at last they prevailed upon them to venture. They were greatly astonished to see such a large kyack, or canoe, as they called us. Some took hold of the ring bolts in the deck; others tried to shake the masts, and in various ways showed their amazement. They remained a short time, saying they were going further up the fiord. They would not land, but appeared to shun the place. Those who were on board also went away in their canoes. The first day the boats were away several fish were seen going up the fiord at great speed. If one rose to the surface it only gave one blast, and the next time it would be a couple of miles off. Our only chance, therefore, lay in their coming near the boats. One day a fish rose to leeward; all the boats set their sails in pursuit. As the wind increased the others gave up the chase, knowing what a long pull it would be to get back. When I wanted to lower my sail we could not manage it, as I had a rope traveller on the mast, the iron one having broken. The former being an old-fashioned plan, and being made of new rope, it stuck to the mast, so that we could not lower the sail.

The wind by this time had increased to a gale, and our only chance of safety was to run before it, and get shelter under an island or any cove in the land which we could find. After running two hours we came to a point of land which opened into a large and beautiful fiord, where it was calm, the water being quite smooth, although in the other fiord it was still blowing, and the sea covered with foam. I have often wondered whether this was the direction the whales took going south, instead of the outside. As we approached a low sandy beach we heard a great shout from a number of natives running about with their bows and arrows. I immediately pulled towards them, but they evidently did not want us to land. We beached the boat, however, near them, saying "chimo, chimo," which means welcome, and is generally their first salutation. They were much surprised to hear me address them in their native tongue, which I could speak very well, being so often amongst them from my boyhood. This tribe had more the appearance of Indians than those further north. I deemed it prudent to fire off my rifle and harpoon gun, fearing some accident might occur, as they are very inquisitive. When they heard the echo of the report among the hills they screamed and yelled in a strange manner. When I first landed, an old woman purloined two pair of mittens out of my jacket pocket. When I attempted to get them from her she fell down, and in consequence the natives appeared very angry. When I told them she was stealing they laughed, so I was very glad to satisfy them by giving her one pair to get the other back. When I fired my rifle off I shot a malemauk, which was sitting on the water. One of the natives took his canoe and fetched it. They were very much surprised to find no arrow, but a hole cut through it, and when I fired off my harpoon gun without killing anything their astonishment was unbounded. The wad, which was made of rope, lay on the water like a ball of oakum. This they fetched on shore, and examined it closely, and they could not understand why a small gun should kill a bird without seeing anything come out of it, and the large one doing no damage but leave a piece of stuff floating on the water.

We were wet and hungry, so quickly made a fire with some coals and wood, which we always carry on such occasions, and made our coffee. It would be an

interesting sight to see us sitting down with fifty or sixty
Esquimaux standing round. We did not entertain the
least fear of them doing us any harm. When we had
finished our refreshment they helped us to haul our boats
above high-water mark. Night coming on, there was no
possibility of us reaching the ship until the wind abated.
We distributed all the small articles belonging the boat
amongst us to take care of. Our new friends then invited
us to enter their tents and partake of their food, consisting
of the stomach of a deer, which they consider a great
delicacy. I refused it with thanks, although they said it
was very agreeable to the taste. It might be to them, but I had
no relish for such dainties. We gladly accepted their
kind hospitality to take shelter with them for the night.

At one o'clock I awoke ; the moon was shining brightly,
the wind had gone down, and I called the crew to launch
the boat and pull for the ship, which we reached at seven
a.m. Only half-an-hour was allowed for breakfast, and we
were sent away again to the other side of the fiord to join
the other boats. No time was given for an explanation for
our absence until the evening. The captain blamed us for
running during the gale. Had the traveller been of iron,
we could have lowered the sail with ease, as the other boats
had done. If we had attempted to bring the boat to, blowing as it was, it would have capsized. The next day we got
under weigh, and towed the ship during a calm to another
harbour or small cove. About two miles further up the
fiord were some natives who had taken up their winter
quarters, and were joined by those with whom we had previously taken shelter, so that there was a large party of
them. We were also in company with Captain Buddington's vessel, the gentleman whom I formerly mentioned.
The following year he wintered in this harbour, and Captain
Hall, of the Polaris expedition of 1871, was passenger with
him in order to become acquainted with the country and

H

habits of the natives. They filled their vessel with whales, intending to sail the next day, but, unfortunately, they were twenty-four hours too late. The harbour froze over during the night, and they were frozen up for eight months, with only six weeks' provisions for a complement of thirty-six men. He afterwards told me his feelings were beyond description at the prospect before him. The natives were chiefly those with whom he was acquainted. He had wintered several times previously, but had been provided for it. One can imagine his feelings of gratitude when the natives came on board, and said, " Here is plenty of walrus ; you shall not starve—we will provide for you." This neighbourhood appears to be the winter home of the walrus. As soon as the ice is strong enough to bear them, they lie upon it, each one making a hole for itself, so that they can go down in search of food. During the winter all hands lived on shore with the natives, dressed like them, and lived as they did. Weather permitting, the crew went with the natives walrus hunting, and, without their kind assistance undoubtedly all would have perished, as none but Esquimaux can hunt walrus in the winter season. Their patience is wonderful. They build a snow wall for shelter from the wind near the holes, and stay watching, without moving, for many hours at a time for the walrus to appear. They kept all the crew supplied with food for eight months without any thought of recompense, and when the vessel was liberated from her long winter's imprisonment, the men were in perfect health and strength. There had not been the least sign of scurvy or sickness amongst them. I agree with Captain Budding-ton, that when scurvy breaks out in a ship's company, nothing will cure it so rapidly as a thorough change of diet, such as raw seal or walrus flesh, with the blood to drink— that is, when it is possible to get it. My opinion is that scurvy arises chiefly from darkness, confinement, and not being able to take sufficient exercise in the daylight, but I leave the question to those more conversant with medical matters. On a previous voyage, Captain Buddington's crew suffered severely from scurvy. He lost thirteen men out of twenty-one. The remaining eight partook of raw seal flesh and blood brought them by natives, and they soon recovered. In this harbour we lay at anchor in forty-five fathoms, and moored with two anchors and sixty fathoms of cable on each, and rode with a heavy cable and swivel with which we were provided. We had only room enough to swing clear of the rocks. Not more than two ships could anchor here. At the head of the cove the land was low, with a sandy beach ; the side cliffs were very high and steep. In the interior was a pond or lake, which the natives said abounded with salmon. Above the low land a brilliant comet was seen to perfection, probably in no other part of the world could it be seen so bright. I believe this comet was visible over all the northern hemisphere. We first saw it when off Home Bay, and I watched it with great interest for four weeks, when the nights were bright and clear. It was a most splendid sight. The tail extended a very long distance. Other stars could be seen distinctly through it, and the heavens and the low part of the land were illuminated by it.

Captain Buddington's small brig was brought up close to the beach, near the encampment of Esquimaux. The other side of the fiord was composed of numerous small islands and sunken rocks, but where we lay the land was steep and precipitous, and the captain gave us strict orders not to pull with the oars when on the look-out for whales, but sail as much as possible, as he thought that the perpendicular rocks attracted the sound of the oars under water, and the whales, being very quick of hearing, would not rise near us, although it is very dangerous to get fast to one with the sail set.

On the morning of the 1st October, as usual, seven boats were sent away ; a fresh breeze was blowing, and we set our sails.   When about seven miles from the ship we cruised about, the weather being bitterly cold, with a nasty sea running.

Contrary to orders, four of the boats parted company from the others, and took shelter under the lee of an island. The second mate was a little ahead of my boat when a whale rose about six feet from him.   He fired at once, and buried the harpoon in its body ; the sail was quickly lowered, and away went the fish to windward at a great speed for a time.   As soon as possible I pulled up to it, and succeeded in striking it with both the gun and hand harpoons.   The whale became furious, and struck about with the tail and fins, rolling over and over near to us. We prepared to lance the first opportunity, and I delivered one in the left side, and was about to give her another, when she struck the boat on the stem, starting it, and leaving some of the skin on the head sheets.

The sea began to increase, with showers of snow, and no time must be lost in trying to kill her, as she was hastening towards the outside of the fiord, as they usually do when attacked in bad weather.   The other boat being further away, and having to pull to windward, only made slow headway.   If the other four had been with us the task would not have been so difficult, and would have prevented me being severely injured.   A boat entangled as we were, is placed in a most dangerous and difficult position to lance. Having no other assistance at hand, I was obliged to use my utmost endeavours to kill the fish regardless of circumstances.   I was in the act of delivering a third lance, but, owing to the rough sea, the boat could not be managed quick enough to get clear of her, and she struck me across the shoulder blades with the tip of her tail, knocking me overboard with the lance in my hand.

My boat's crew afterwards told me that immediately I was struck by the fish it began to blow blood, and went rapidly

to windward. The line had got foul round the bollard of
the boat, and the axe could not be found to chop it, having
got displaced when she struck me. The danger was
imminent, for if the whale had gone downward she would
have carried the boat with her. As soon as possible the
line was cut with a knife, and the boat pulled towards me.
I was lying on my back like a cork on the water. When
they took hold of me the lance fell out of my hand. I was
lifted into the boat, conveyed on board, and put into bed.
My clothes had to be cut off, being frozen. Fortunately
we had a very clever doctor, who waited and watched by my
side for six weeks until we arrived home.

To return to the incident of the whale. The other four
boats were told by the American boat's crew that we were
fastened to a whale. They made all speed towards us, but
did not arrive until she was dead. There was a short heavy
sea, and it was with difficulty they could lash the fins
together and tow her to a place of shelter. One boat
was thrown by the sea over the carcase and capsized,
nearly drowning the crew. After towing some time they
found refuge in a small cove about six miles from the
ship, and there stayed until the following day, when the
wind moderated. The boats came alongside during the
afternoon, their crews being wet, hungry, and nearly
worn out. After a rest they "flensed" the fish, and
stowed the blubber away in the tanks. This whale yielded
nineteen and a-half tons of oil and nearly one ton of bone.
For a few days the crews went away with the boats as usual,
and saw many whales, but the weather became so boisterous
they did not succeed in capturing any more. The captain
heard from some natives that there was a harbour about
twenty miles further south, close to Frobisher Straits, and
that whales were plentiful. They got under weigh on the
10th of October, but only reached the outside when night
came on, the wind having failed them, so they lay to and

waited for daylight. During the night the wind rose, and
began to blow with thick snow. All hands were on deck,
and several times the ship was close to some breakers. It was
a most anxious time for everybody, especially the captain, for
we were in an unknown place amongst many sunken rocks.
However, Providence guided us, and when daylight
appeared, the snow had ceased, although the wind was
blowing strong. We found ourselves a few miles from the
land, and the captain decided to bear up for home, to the
great relief of the whole ship's company. When the ship
was headed to the eastward, a line of breakers was seen
ahead. She was hauled to the northward for a short time,
and that was the last view we took of Nugumut. We had
a stormy passage until the Butt of Lewis was sighted, and
then followed a few fine days, which enabled me to be
removed on deck for a short time each day. It gave me great
pleasure to be once more on deck and look around, after
being cramped in a close berth such a long time.

In those days the accommodation was very poor compared
with the present time. The furniture consisted of a table
in the centre of the small cabin, the transome forming the
after seat. The captain's chest was the starboard seat,
the mate's on the port. Such were the comforts of our
cabin. The sleeping berths were very small, so that only
one person could dress at a time, every available space
being taken up for stores, etc.

When we reached the North Sea the weather was very
cold and damp, and being too venturesome I caught a
severe cold that affected my lungs, which had been very
much contused by the stroke from the whale's tail. On
arriving home I was shocked to find my dear mother had
died about six months previously, which was a great grief to
me in my weak state. I was carefully nursed, and had the
best medical advice. I slowly recovered, and by the middle
of January was able to walk out. My medical advisers

thought I ought to go to a warm climate, but against their wishes and those of my friends, I engaged as mate of the brig Anne, bound for the seal and whale fisheries at Greenland. The bracing atmosphere of that climate soon restored me to my usual health and strength.

CHAPTER XII.

A NEW VENTURE—COUNTING CHICKENS BEFORE THEY ARE
HATCHED—TAKING THE PACK—GETTING STOVE—DEATH
OF THE DOCTOR—1st OF MAY PROCEEDINGS—ENCOUNTERS
WITH BEARS—TESTING A BOMB—LANCE ON A FINNER
WHALE.

IN 1859, having recovered from the afore-named accident,
we fitted out at the same time as the Hull whaling
and sealing fleet, and proceeded to Shetland, and lay in
Lerwick harbour waiting for men to make up our crew,
which was rather difficult on account of the unusually large
fleet lying in the harbour, also bound for the Greenland
sealing, and requiring men.

One steamer specially attracted much attention. She was
called the Empress of India, belonging to Peterhead, built
of iron, and fitted out expressly for the trade She was
strongly fortified, being twelve feet thick forward, and
carried eleven boats. The bottom of the captain's gig was
bronzed, which made it look very showy. No expense had
been spared in her outfit, and she was manned by 110 men.
All the crew expected they would make a small fortune, and
looked upon our sailing vessels with contempt. Some of
the officers were so sure of getting full of seals that they
made all their plans for the future, viz., they were going to
fall in with the north end of the body of seals and sweep
through the centre, leaving the rest for those who were so
fortunate as to be in their company. But alas, as the
Scotch say, " Schemes of mice and men gang aft agley."
The first piece of heavy ice they came to, they struck it with

the bluff of the port bow, which stove her so that she
foundered in four hours. All hands were saved by the
ships they had so recently ignored. But what a serious loss
to the enterprising owners. Several iron steamers belonging
to Hull, called the Emeline, Gertrude, Corkscrew, Labuan,
and the Wildfire, a wooden sailing vessel, were fortified and
proceeded to the seal fishery. Most of them came back
empty and damaged. One called the Labuan captured
between eight and ten thousand seals. This year proved that
iron steamers, however strongly built, were not suitable
vessels to contend with the Greenland pack. The winds
prevailed from the eastward, and blew heavy at times,
which caused the sailing vessels to be wary, and keep
off the heavy tight-packed ice. After one easterly
gale had subsided, and the wind changed to the
westward, we worked the ship towards the edge of
the ice to the northward of Jan Mayen's Island. We
had an idea that the body of seals were in that
direction inside of the ice, which proved to be the
case. On arriving at the pack edge at the bottom of the
bight, the ice was still tight. We dodged about, waiting
for it to slack, but it was otherwise ordained, for suddenly
the wind veered to the eastward and commenced to blow.
We carried all possible sail to beat off the lee ice, but
could not do so. The gale increased, so we had no other
alternative than to take the pack, which is a last resource.
All hands were called to attend the ship. The master was
at the mast head looking for the best place for this purpose.
In taking the pack in a gale of wind the scene is most
terrible, and makes many a stout heart quail for what
may be their fate in a short time. The roaring of the gale,
the dreaded white breakers on the heavy ice, and the
anxious looks of all on board as they hear the ice crashing,
make all aware that the first piece they strike might knock
the ship's bows in, when there would be little chance

of being saved, which is very depressing. Yet those in charge of the vessel must put such thoughts aside, and think only of the arduous task before them. When once safely inside, they have time to reflect and thank God for preserving them. The sails were trimmed, and the ship kept before the wind, heading for the pack. Luckily we succeeded in getting inside of the outer edge, or crust, as it is sometimes called. We again set all sail possible to enable us to bore further into the pack, clear of the swell. For a short time our hopes were raised, thinking we might get among the seals and obtain a good voyage. Suddenly the ship collided with an extra large piece of ice, which caused her to broach to, broadside on to a sharp point of it, and made a large hole abreast the main mast, breaking three timbers and staving a cask in the hold. The pumps were at once manned, and in a short time the rush of water was stayed by swabs and other means. The carpenter and some of the crew went below to strengthen the ship's side with shores, etc. In course of time she slipped past the dangerous obstacle, and we got her again before the wind and swell. In two hours' time we were in safety, and lay comfortably. As soon as possible a patch was bolted on the outside, and the broken timbers made secure by spars in the inside to the mainmast, which was opposite the damaged place. This was a most serious misfortune for us, and caused much anxiety the whole of the voyage.

One beautiful, clear, frosty, bitterly cold morning, we were plying among streams of ice to the northward. I was in my berth, which was shared by the doctor, a very clever, courteous gentleman, who awoke me muttering incoherently. On getting up I found him delirious. I called the captain, who had him taken into the cabin, and made comfortable on the transome locker. Hē appeared very drowsy, but recovered a little after breakfast, and again relapsed into a stupor, from which we could not rouse him, although we

used our utmost endeavours. We hailed several ships, but
none carried a medical man, all being foreigners. The
doctor had been ailing a few days, but had not complained,
and we could not account for his sudden illness. About
ten p.m. we were summoned by the man attending to him,
when we saw a great change had taken place. He had not
spoken since breakfast. I took hold of his hand, when he
gave one sad look, and quietly passed away. As may be
imagined, our breakfast next morning was taken in silence.
One whom we all respected lay lifeless alongside the cabin
table.

About eleven a.m. we met with a Scotch brig, and asked
their doctor to come on board to see the corpse, and give
his opinion of the cause of death, which he did, and made
out a certificate accordingly. The body was carried on
deck, and in a very short time it was frozen like a marble
statue. A coffin was made, in which the body was placed
and packed with sawdust. In this manner we kept it on
board four and a half months, and buried him in the
churchyard at Lerwick.

After the doctor's death, we cruised about among streams
of ice, but found no traces of the seals. The time was
fast approaching when the young seals would take the water
and provide for themselves. One fine day the main pack
opened, and we thought that the body of seals was in the
vicinity, so we set all canvas, and bored into it. We saw
several upon the ice to the westward, and were delighted at
the prospect. But before midnight the pack closed, and
the vessel was immovable. The following morning, from
the masthead, a brig was seen further inside, apparently
sealing. We called all hands, and travelled in that
direction. But when about five miles from the ship, we
were disheartened to find our way cut off by a lane of
water, about a ship's length wide, extending north and
south. We walked a long distance to the northward, in

hopes of getting round it, but were disappointed, so had to return on board, as night was coming on.

Next morning the weather changed for the worse, and came to blow and snow heavily. But it made no movement in the ice. After blowing thirty-six hours, we were fastened into a solid floe without any water to be seen from the masthead. The brig before mentioned was nowhere in sight when the weather cleared. We had evidently drifted in different directions, but afterwards we found out it was the Polar Star, belonging to Peterhead, which was among the seals at the time when we first saw him. It was discouraging for us to be, as the old saying is, "So near and yet so far."

We remained fast for six weeks. Every day, when the weather permitted, all hands were sent away in different directions in search of seals. Some days a few would be brought on board, on others none. Bears were sometimes numerous, but as their appetites were appeased by plenty of food in the neighbourhood, they did not trouble us much. Each officer carried a rifle to protect his boat's crew, if required. Several narrow escapes were recorded. On one occasion, a party consisting of eight men, with two rifles, were imprisoned on a piece of ice for two hours by a large bear. Their rifles had got wet, and would not go off. Bruin walked round and round all the time, which appeared an age to them. It was an unpleasant position for those who were there, but laughable for those who were not. The animal at last took a long look at them, apparently to decide whether they would be tender eating, and probably concluding that they would be too tough, he wended his way to better pastures.

Another time, two men and myself straggled some distance from the others. We killed three seals, and were dragging them to the ship, which was about five miles away. On looking back, we saw two bears coming full gallop towards us. The men were inclined to run, but what use would it have been when we were such a long distance from a place of refuge? We waited a moment to see what were their intentions. I knew they could not be hungry, yet I did not care for their playful manner. On they came, and suddenly stopped about a ship's length from us, and snuffed at the light track of blood. I then thought they would go away, so we again began to travel forward. In this way they continued for some time, and at last we took no further notice, but walked on. If they had come much nearer we should have left our seals to occupy them. My reason for not firing was, I could not distinguish the male, and if I had shot the female, the other would have been doubly furious, and have been upon us before I had time to reload. It is seldom two males travel together. They are unsociable brutes, and prefer solitude to each other's company.

When we arrived on board, the captain said he was anxiously watching us from the crow's nest, for they appeared to him to be alongside of us. We had their company within a mile of the ship, when they retraced their way back, we being heartily glad to part with their society.

Several bears were shot, and if it had not been for such excursions our time would have passed very monotonously. When the weather was very clear fifty-two sail could be counted from the masthead. At another time not more than a dozen. Our ship's head did not vary more than three points during the six weeks we were fast, and no crack was perceived in the ice to change the position of the other ships.

One day three men travelled to us from a brig lying about seven miles away; they had been as near to the seals as we were when the gale came on. From them we heard that a foreign brig lying next to them had a man eaten by bears a few nights previous. It is foolish to go alone from the ship when there are so many dangers around. All that remained of the poor fellow was his boots, a few bones, and torn clothing.

The 1st of May was ushered in with a fresh breeze and thick snow. All preparations had been made to welcome Neptune and his friends on board, similar to what is practised when crossing the line. Troubles were for the time forgotten. The between decks were cleared, play bills posted, judge and jury boxes erected, and the barber's shop made ready for greenhorns, *i.e.*, young men who had not been to the Arctic regions before. At midnight eight bells were struck. A gruff voice hailed the ship through a speaking trumpet, demanding the name of the vessel, the captain's name, the luck we had had, etc. He concluded by asking if we had any of his children on board, meaning *green hands*, who were at the time stowed in the fore peak amongst the coals, with a watchman guarding them.

When Neptune's questions were answered satisfactorily, he was politely requested to come on board. The scene was most amusing. His carriage was a main hatch, and his majesty was accompanied by Mrs. Neptune, both in dresses made from rush mats, shavings for curls, and a potatoe net for a veil. Their faces were coloured red to make them look fresh.

They were followed by their retinue, comprising the barber, similarly dressed, but with a huge paper collar, stiffened with white paint. Next came the barber's clerk, policemen, and jurymen. These carried the music, consisting of tin pots, pan lids, and other noisy gear. The procession went round the deck, halted at the cabin door, sung a song waiting for an invitation to go down, or orders from the captain for the steward to give each a dram. They expressed their thanks for the same, and good wishes for a prosperous voyage, and retired to the between decks, when business commenced.

The policemen wore oil jackets, with large buttons
painted white. Seats were taken, and the poor fellows
brought out singly. Neptune enquired his name and
occupation. He was then placed on a block of wood, and
the barber commenced to operate upon him. The first
article shewn was a formidable razor, made out of a piece
of iron hooping. The blade was eighteen inches long, with
huge gaps in the edge. The poor innocent was then
lathered with coal tar, and his hair powdered with crushed
chalk and resin, well rubbed in. This was for those who
had not made themselves agreeable, or had a tendency to
be idle. Those who took the farce good-naturedly got off
the best. It was no use being refractory, as everybody
taking part in the programme were bound to assist each
other. When the shaving was over, and a few songs sung,
things were restored to their former places, and I am happy
to say throughout this formal custom everything passed off
without any disagreement.

The garland was placed midway on the main top-gallant
stay by the youngest married man. It was fancifully
decorated with ribbons, presents by sweethearts and friends
before sailing, for that occasion. Garlands have different
devices in the centre, such as ships, crowns, hearts, etc.
They remain hanging until the vessels arrive home, when
there was a scramble by the Trinity House boys and others
who should get the coveted prize, now bleached white with
the weather. It was considered an honour to gain
possession of one. I have heard old men speak with
delight of having secured such a favourite ship's garland
when they were boys. No doubt there was a history
attached in former days to these strange customs.

Three weeks after this a heavy gale sprang up from the
eastward, which broke up the frozen pack and liberated us.
The wind veering to the southward enabled us to reach the
outer edge. There was a very heavy swell among the ice,

which made it a most anxious time for us all, on account of our damaged condition, but with great care we got into the open water without further mishap, although we received some very heavy thumps on our bows.

We then proceeded to the northward, amongst straggling streams of ice, in search of seals, which make their way in that direction in separate bodies. Sometimes good voyages have been made when the weather was fine by shooting over the streams of ice, which are covered with seals. At times, after the first shot, all disappear, yet at other times hundreds have been shot in one day. The season was getting late for them, as we had been frozen up so long.

Troubles seemed to follow us quickly. Many men were on the sick list, especially the carpenter, who had been ill several times during the voyage, but the doctor had always pulled him through. He now appeared worse than ever, so we slung a hammock in the cabin, where we could better attend to him. He expressed a wish to be sent home. A Scotch brig in our company was going home, and the captain kindly consented to take him. The poor man was so delighted that he dressed himself and stepped into the boat quite cheerfully, and to all appearance had taken a new lease of life. He died a few days afterwards, and was buried at Lerwick, where we afterwards laid the doctor.

Proceeding further north, we came to the whaling grounds situated off the island of Bontekoe and Cape Broer Ruys, among very heavy loose floes and sconces. This ice sometimes attains the thickness of thirty-two feet. We sailed and dodged about in search of whales, and saw but few, although the ice lay in such a favourable position for them. The fogs being so frequent prevented us from sailing about as we otherwise should have done.

One day we came up to a brig which had found a dead whale. The carcase was extended to a great size. The crew had been busy for some time shooting bears which had taken charge of the body, and were very reluctant to relinquish it. Before we arrived they had shot thirty-two. There must have been over a hundred on the floe when we came up with our boats. We shot four in the water, and many lay sleeping further away, evidently having had a good meal. These animals disputed the prize. They stood upon it and swam round it in defiance for two hours. Bears are keen-scented animals, and no doubt had scented the dead whale many miles off, which must have been the case to attract such a number to one place. Very little blubber was on the carcase ; sharks had eaten all below the water, and the bears all above. The whalebone was taken out, and the body again left for the benefit of the sharks and bears. A few days later our vessel was made fast to the land floe, and I went away with my boat and crew to shoot birds. We were also prepared if a whale or narwhal (sea unicorn) came in our way. As we were pulling close to the floe, which was six feet above the water edge, a large finner rose close to us. I drew the harpoon and wad out of the gun, and replaced them with a bomb lance. These lances were of very recent date, and had only been in use about a year. Not having tried one before, I thought this was a good opportunity to prove the efficiency of them. The finner, which was about one hundred feet long, was only fifty feet from us, and formed a barrier between us and the outside. I did not think of our position, being too intent in trying what effect the bomb lance would produce. We were so well situated that I could choose to strike it in any part of the body, so took aim behind the left fin and fired. I then perceived our critical position. If it had rolled towards us instead of from us we should have been crushed, as there were no means of escaping on the ice. Immediately the bomb entered the body the animal appeared paralyzed, and in a few seconds the bomb exploded, and we felt the vibration. Then the body appeared to expand, and rolled

I

from us in agony, giving a great flourish with its fins and tail, causing the water to nearly fill the boat, and wetting us to the skin. The next we saw of her was about a mile away, blowing blood, and swimming at a rapid rate through the water. The boats from the ship were in pursuit. The finner whale we never entangle with harpoons, on account of their great strength and endurance. They would take away all our lines. The blubber is of a glutinous nature, and of little value. The whalebone is short and very coarse, so they are not worth the risk. All our endeavours to make a payable voyage proved fruitless, and the evenings becoming dark, we could not remain any longer, so bore up for home with a fair wind.

One hundred and fifty miles north of the Shetland Islands we came though large shoals of herrings making their way south, and on arriving at the north island, called Unst, some fishing boats came off to enquire if we had seen them. On giving them a reply in the affirmative, they were satisfied that they would soon begin to reap their harvest, as the herring fishery is one of their chief industries. I have often wondered where the herrings came from, as I have never seen any in the Arctic regions ; yet, when we passed through them, they were coming direct from the north. Soon afterwards, we brought up in Lerwick harbour, and took the body of our doctor on shore and buried him not far from the grave of our late carpenter. A fair wind springing up, we got under weigh, and in a few days once more arrived at Hull, and moored in the dock.

This had been an anxious and disheartening voyage, after using our utmost endeavours to make a profitable one for ourselves and owners. We had very little money to take ; in fact, most of the harpooners were in debt to the captain, thus taking away part of the pleasure of coming home to our friends. Such is sometimes the lot of those who venture on these speculative voyages.

CHAPTER XIII.

HARD TIMES—DROOPING SPIRITS—A FLOATING ISLAND—
BROOMING—MIRAGES—ONCE MORE NEARLY TRAPPED.

NEXT year we once more fitted out the same brig for
Davis's Straits, calling, as usual, at Lerwick, and left
with the first fair wind for the whaling grounds.

A fresh breeze prevailed from the eastward, which made
us imagine we should speedily arrive at our destination. In
a few days the wind changed, and blew furiously from the
westward, driving us back many miles. We were under
close-reefed topsails for four days. After struggling for a
long time, we made the ice off Sukkertop (Sugar loaf), the
land being forty-five miles distant.

After much detention we arrived in the neighbourhood of
Disco. Several whales were seen, and the boats sent in
pursuit, but without success. Every advantage was now
taken to get north. We cruised about for some time
on the whaling grounds off Black Hook, subsequently
coming to a place called Proven, a settlement to the southward of Sanderson's Hope, which is a prominent headland,
upwards of 3000 feet high. There is a large loomery or
breeding-place for the great auk, called by the sailors, looms.

The natives belonging to Upernavik have a long rope
attached to the cliff, and during the season gather quantities
of eggs in a manner similar to the people of St. Kilda in the
Lewis Islands, and some parts of Shetland. We made fast
to an iceberg under the lee of an island, close to the mainland, and were soon joined by the s.s. Lady Seale, belonging
to the same owner, which made fast alongside, and we dis-

charged fifteen tons of coals into her, which we had brought
out in expectation that she would tow us when required.

In a short time the ice slacked, and away she steamed,
leaving us to the dreary work of towing by our boats.
More sailing vessels came up, which made it more lively,
as we had been struggling along by ourselves, and taking
every advantage of the least slacking of the ice.   I remarked
that the crews of sailing ships did not tow or track so cheerfully as formerly.   I came to the conclusion that since steam
was introduced the spirits of the men drooped; and no
wonder, for the poor fellows would be towing at the rate of
one and a-half or two miles per hour.   A steamer would pass
them at the rate of eight miles, with their crews leaning over
the bulwark taking their ease ; but one must endeavour to
look at the bright side, and do our utmost to secure a
remunerative voyage.   We threaded our way through the
intricate passages amongst the rocks and small islands close
inshore, until we arrived at the Duck Islands, but got no
further.   From the masthead, as far as the glass would carry,
the floe was broken up inshore, and no land ice to be seen.
This is one of the greatest drawbacks in going through
Melville Bay.   No man would run so great a risk of getting
beset amongst the loose floes, especially with a sailing ship,
as there is no knowing when they would get liberated.
With steam there is a probability of forcing their way.

We waited some time, and seeing no possibility of getting
across the bay, we retraced our way back to the southward,
and searched everywhere for an opening.

This year was noted for the prevalence of dense fogs,
which impeded our progress.   Once we had a fog which
lasted six days, and knowing we were some distance from
the south lowland on the west side, north of Cape Hooper,
the officer whose watch it was on deck called down the
cabin that the vessel was close to the land.   The ship was
immediately put about, and a boat lowered.   We could not

account for being so near, as by our calculation we ought to be forty miles from it. Taking a gun with me, I pulled towards the supposed land, and found it to be a large sconce of heavy ice, covered with gravel, sand, and large stones, some of which would weigh upwards of a ton. This piece of ice must have been attached to the land under a perpendicular cliff. This may account for large stones being carried hundreds of miles away during the glacial period.

On my return I reported what it was to the captain, who considered it best to make fast to it until the weather cleared, when we could ascertain our position. When all was secure, the crew were soon off for a run, and found a large pool of fresh water. All hands were soon employed filling up our fresh water casks, as we had been compelled to melt berg ice for our use. This large piece of ice, or as it might be termed a floating island, was about one mile in circumference, and twenty-four feet thick. Shortly after we had got our fresh water on board the weather cleared, and the west land was in sight about thirty-five miles away.

We cast off, and began to work our way to the S.W., amongst the ice, which was slack in that direction, and reached the west water. Some land ice lay between Cape Broughton and Brodie Bay. We ran along it for a few miles, and fell in with four or five ships which had crossed over to the northward of us. They also had been detained by the fog. In a few days we got as far as Cape Kater. A fine breeze had set the loose ice off. There we found four more vessels made fast to the land floe. One of them was killing a whale. We thought surely we should have an opportunity of securing one or two, but before we were able to get to the floe a fresh breeze sprang up from the eastward, and began to break it up. In a short time it was a pack, and the ship which had killed the whale was tight beset, and was with great difficulty liberated a few days later. This forced us to work our vessels further south in more open water, lest we should get beset. We spoke several ships, which, like ourselves, were not successful.

I will here give a description of the term brooming. When two vessels meet, and the masters are conversing with each other from the crow's nest, the men from the deck signal in the following manner to know how many fish each ship has got : One waves a besom or birch broom three times, then holds it upright a moment, then makes a stroke downwards for each whale they have captured, and then gives a final flourish. The other replies in the same way.

Formerly many whales' jawbones were brought for farmers, who used them for gate posts, and they may still be seen in many parts in good preservation. Of late years they do not pay for the care and trouble of bringing them home. The price used to be 30s. per pair. Sometimes when a ship got full they would hang a whale's tail across the stern that the constant dropping of the oil from it would break the heavy seas when running before a westerly gale. The use of oil on the troubled waters appears to have been an old custom amongst the whalers.

The mirage during the latter part of July and August was very remarkable. On one occasion we sighted a ship which was thirty miles distant. It was distinctly visible with the naked eye from the mast head, and appeared in its natural position, also inverted in the air with the masts touching. It was an interesting and singular sight. This phenomenon I have seen but twice during seventeen years.

Another time I saw an iceberg at a distance of forty-five miles, changing into many peculiar forms ; this is a frequent occurrence. I have also seen the east and west land both at one time, between Holsteinberg and Cape Dyer, latitude 66½° N. The distance from each other is 184 miles, though neither of those places are very high. In Melville Bay especially the mirage is very deceiving. One can fancy seeing a body of water distinctly, with pieces of ice and bergs in it. Later on it appears to be a solid field of ice; at last it proves to be a small hole of water not exceeding a quarter of a mile in length. Frequently it has been reported from the mast head that a body of water is seen to the northward, and immediately afterwards all ice. Ships at times a short distance away have very fantastical appearances. Once in particular, three of us were not more than a mile apart, with all our boats towing. Each one could distinctly hear the other singing, and distinguish the words. At times the hulls would appear to be short blocks of wood, with the masts high in the air. The boats had a similar appearance, with their men above them. Quickly a change would take place, the ships and boats were elongated, the masts and men looked like specks. The same effect was also produced on icebergs and other objects.

To return to our narrative. The main body of ice continued to drive gradually south. Our only chance was to reach Cape Hooper and anchor in the harbour, with several other ships. We sent our boats away as usual to the outside.

One day a vessel came in bringing part of the crew of the s.s. Chase, belonging to Hull, which was crushed by the ice and beached inside of Button Point, on the north side of Pond's Bay, to try and repair her. This was the only resource left for them, as the vessel would have sank if she had not been put ashore. Unfortunately a strong southerly wind sprang up, which caused the large floes to drive her further on the land, and she became a total wreck. The crew escaped in their boats to the other steamers in the offing.

The ice now began to come into the harbour. Some of the sailing ships got under weigh and proceeded to the outside, as the body of ice in the offing was gradually closing on the land.

Our little company consisted of two steamers and two sailing vessels, which stayed in the expectation of the wind drawing off the land, and a probable chance of securing a whale or two. This is all very well, if we knew for a certainty that it would be so ; but having been caught here once before I was very dubious. When these inlets and small bays are filled with ice in the latter part of the year, it seldom clears out until the next autumn. Instead of the wind changing, it blew stronger from the north, bringing a greater body of ice down, and cutting off our way out. The prospect was a very dreary one, we having no extra provisions. However, the steamers which were in our company kindly took us in tow, and with great difficulty steamed with us to the outside, where we lay until daybreak, when they again took us in tow until we reached clear water, which enabled us to work under canvas.

The ice continued to drift rapidly south in a body, and pack along the land ; this prevented us from fishing. The weather increasing in severity, we could not lower our boats for several days on account of the heavy sea, and there was no possibility of getting into a safe harbour, unless we ran to Cumberland Gulf, but it was too late in the season to go there with a sailing ship. The most prudent course was to bear up for home. We had a good supply of fresh water, which we had obtained in Cape Hooper harbour. After stowing our boats, we bade adieu once more to the country of ice and snow.

During the passage home we encountered heavy weather, which drove us northward. The first land we sighted was Suderoe, one of the Faroe Islands, and the wind still being southerly, we stood close to the land and tacked. The next day we were able to weather Monk Rock, which lies to the southward of Suderoe, at the extreme end of the Faroe Islands. This rock was a high narrow pinnacle, resembling a craft under canvas, similar to Rockall with one exception, the former being brown—the latter of a white appearance. The Monk Rock has now disappeared. A few days afterwards we arrived at Lerwick, discharged our Shetland men, and with the first fair wind started for Hull, thus ending another voyage.

CHAPTER XIV.

THE following year—1861—I was appointed master of my old ship, Truelove.

We left Hull the first week in March, and brought up in Grimsby Roads on account of a strong south-easterly wind blowing. We lay at anchor two days, when a favourable wind sprang up, then got under weigh, discharged our pilot, and proceeded towards the Shetland Islands.

Passing Sumbro' Head, we encountered a strong wind with thick drizzling rain, and eventually arrived safely in Lerwick harbour. We remained there a few days, and shipped our complement of men ; then set sail with a fresh southerly wind through the North Channel. After passing the island of Unst, we hauled to the westward. An easterly breeze favoured us for three days, enabling us to make our southing, and bring us into the usual track.

During our passage towards Cape Farewell, we fell in with many icebergs in Lat. 58° N., Long. 44° 10 W. The day was beautiful and clear, and the clouds near the horizon to the northward appeared so much like the land with its snow-capped mountains that any experienced person might easily be deceived, although we knew the land to be about one hundred and ten miles distant. When such clouds appear, they are called "Cape Flyaway" by the sailors. Three other ships were in sight, which broke the monotony, and made it more cheerful for us, after the stormy weather we had encountered.

At midnight the barometer suddenly fell very low. The Aurora Borealis was very brilliant, and we were surrounded by coloured vapours. This phenomenon was followed by a very heavy northerly swell, which warned us to prepare for a gale. The sails were securely stowed, except a close-reefed main topsail. Contrary to expectation, it remained calm two days. The tremendous swell caused the ship to roll with the rails under, which made it most uncomfortable. No one could rest in the berths, so hammocks were eagerly sought for. In those days each man provided himself with one. Although we were not caught in the gale, it was very near us.

The master of one of the ships which we saw afterwards, told me they had encountered a terrific storm, and had a narrow escape of being in collision with an iceberg. To use his own expression, the wind howled like ten thousand cats. The distance between us at the time was only thirty miles. During the day he was a few miles north of us, and had a breeze which carried him into the gale. Moderate winds prevailed four or five days, enabling us to get further to the westward, and we then met with a very heavy gale, which made us lay to under close-reefed main topsail and balance-reefed main trysail. The main staysail was reefed, and a double sheet rove ; the halyards through a leading block ready for immediate use in case of emergency in heavy weather, especially amongst bergs. This precaution proved our safeguard on the first night of the gale—a night to be remembered by everyone who was on deck. The wind was blowing furiously, and the sea running very high, and not a star to be seen in the inky black sky. Thick sleet fell, and the waves broke over our little barque. The watch on deck and myself were lashed in case a heavy sea struck us. An iceberg suddenly appeared close under our lee, and there was not a moment to lose, in face of the great danger which threatened us. Up went the main staysail; the ship answered her helm splendidly, and, under the guidance of Providence, we passed the berg by a hair's breadth, the back wash helping to carry us from it, and in five minutes we lost sight of it in the darkness. Our hearts nearly stopped beating; we scarcely could breathe, so intense was the agony of suspense; and, after the danger was passed, no one spoke for a short time. At last a man said, "That was a close shave." The main staysail was again hauled down, and all were thankful we had escaped the danger.

A strict look out is always kept, especially on arriving in longitude 30° west. One hand is sent on the fore yard, and relieved every hour. This duty is always given to an experienced man, who has orders to report anything which may appear like a berg. Sometimes a sea breaking white may be taken for one, and the ship's course altered immediately. But no blame is attached to the look-out if it proves not to be ice. It shews he is diligent in doing his duty.

I may relate an incident that occurred, which taught a lesson to everyone visiting Davis's Straits. Many years ago a ship called the Shannon, belonging to Hull, struck a small berg, which knocked her bows in and killed some men below. The ship turned over, but the empty casks in the hold kept her from sinking. Some of the crew were drowned. Those who were saved were on the bottom of the ship two days, and were providentially rescued by a Danish vessel, which was taking out stores for the settlements in Davis's Straits. After this shocking casualty an extra good look-out has been kept for bergs and ice much sooner than in previous years.

After the gale we had a calm, so got our rigging and other damage repaired. We made the ice south of

Sukkertoppen, and worked our way to the northward amongst streams and patches, until we came to Whale Fish Island. The natives came off and told us that many fish had recently been there, so we coiled the lines in the boats and prepared for work, cruising between the Islands and Disco. Many whales were seen and fired at, but missed. Such mismanagement was very disheartening at the commencement of the voyage.

We stayed two days longer, and not seeing any more, we sailed to the westward amongst streams of ice, and to all appearance this was the most likely place to find them. I thought possibly we should get some here, but was disappointed, although in previous years I had seen them very numerous. We then sailed further north to Disco Bay, but the weather was too boisterous. We fell in with some of the other whalers, but none had been successful, and the season for the Disco fishing was over.

We all turned our attention towards N.E. Bay. When off Hare Island a strong gale from the northward commenced to blow, and before we could get made fast to an iceberg the ice wrapped round our ship and two others, and we were soon beset. The ice crushed and pressed upon us in such an alarming manner that we expected every moment to become total wrecks. Our rudder was promptly unshipped, or we should have lost it.

In the offing of this island there is a reef on which bergs ground, and the ice drifting from the northward upon them, makes it very dangerous to get beset, so we always hug the land if possible, and take the inside passage, called the Malygat, but this time it was full of ice, and our only way was at the outside. One of the vessels was forced close to a wall-sided berg, and had scarcely time to take the boats in before the ship had drifted within three feet of it.

In our case nothing but a small sconce piece of ice saved us from being forced on to another. Fortunately the floe split ahead, which altered our position, and we drifted past it. This was a most critical time for us all, and proves that sailing ships are safest when holding to the fast ice. Steamers can easily liberate themselves when the pack slacks, but we are at the mercy of the winds and currents.

The gale lasted twenty hours, and gradually calmed. When we attempted by warping to get into a hole of water which made its appearance in Malygat Strait, we made little progress, as the ice was so tight. To our astonishment a passage suddenly opened close to us, in the direction of Four Island Point. The rudder was immediately shipped, and the boats commenced towing inshore. After four hours' work we made fast to a small berg close to the land, not being able to get any further, I went on shore to have a look from the hill top. There was a large body of water to the northward, and no vessels to be seen except the Æolus and Emma, belonging to Hull, which were still beset. A northerly wind sprang up, and a strong current began to run southward, which caused the loose ice to pack upon the land, and prevented us getting round the Point. Fortunately for the Æolus and Emma this liberated them, and away they worked northward with their ensigns flying.

We had previously pitied them, but our situations were now reversed; such is the sudden change of fortune to which we are subject in this country. When expressing sympathy for others we may in a short time need theirs. We could not move from here until the ice had drifted past. A change came at last which liberated us, and we set all sail and plied to the northward.

When off Black Hook, we joined four other ships, and at Upernavik we saw the other vessels about five miles north of us. The weather being calm, all hands were employed towing, and we got within three miles of them. They were waiting for a narrow neck of ice to open. A light breeze sprang up ahead, which prevented us towing in that direction, and seeing an opening between two islands more inshore, we set the fore and aft canvas, expecting this passage would bring us to the northward of the other fleet.

Here we met with another disappointment, in finding a piece of land ice extending from one island to the other. It was not more than a quarter of a mile wide, but it prevented our progress. Our retreat was also cut off, as the loose ice had come down ; so we were obliged to make fast to icebergs.

This neighbourhood consists of numerous small islands and sunken rocks, extending far from the mainland. Our little fleet was composed of six ships. We were close to the large glaciers at a place called Ankapadluk. We saw some massive bergs break off, which made a heavy swell, and broke our warps.

The detention at this place was the cause of us not getting a full ship, and the loss of two vessels soon after in Melville Bay. A light breeze ultimately springing up from the southward, broke away the barrier of ice. We cast off from the berg, and ran through the channel between the islands. The water was so shallow that the rocks could plainly be seen under the ship's bottom as we sailed along. We had passed the most dangerous part in safety, when the wind increased, with thick snow. We were obliged to run and take the risk, as there was no place of shelter. Two large icebergs, scarcely a ship's length apart, appeared ahead, and the only alternative was to run between them.

We were no sooner through than the bergs closed with a terrible crash, splintering huge masses from their sides. Luckily, we did not perceive them closing until they were passed. One must have been aground and the other afloat. It is surprising how those massive pieces of ice travel amongst such broken ground as there is here. Large glaciers are in this neighbourhood, and it is the most rocky part in Davis's Straits.

After grappling our way amongst numerous icebergs and small islands, we came close to Berry's Island, which is the outer one of the group. The weather cleared but the wind was still blowing. There was no possibility of getting through the floes, which had closed within the last few hours, but we could plainly see the water to the northward, and it was with heavy hearts we were obliged to make fast to an iceberg close to the island. The wind was still strong from the S.W., but the weather was clear, and we knew the northern fleet would be making the best of their way towards Melville Bay. The wind again increased in force, with thick snow, but we were sheltered by a reef of bergs which broke all pressure from the fast drifting floes. This weather continued for twelve hours, when the wind fell and it became calm and clear. The floes in the offing were still drifting northward, shewing that there must have been much water in that direction. We hourly expected an opening in the ice, but this did not occur until the wind changed to the N.E., but it did not slacken the ice near us until the following day. Owing to the wind being ahead, we were obliged to warp from one berg to another, which was a very tedious process.

We ran out a boat's lines and sometimes more, according to the distance between them, and tracked along the deck in this manner. After warping ten hours, we got into clear water, set our sails and plied northward. After passing the Duck Islands, we came in sight of the Devil's Thumb, and made fast to the land floe.

A few miles to the westward there was a deep bight, and we held a consultation on board one of the ships which would be the most prudent course to take, so we concluded to remain by the fast ice. Caution and prudence are always necessary, but cannot at all times be put into practice. In this case, if we had gone amongst the loose floes to the westward we could not have been worse than we were, as it afterwards proved.

In a short time the whole Bay broke up into floes, and no fast ice remained. A calm, prevailing for a few hours, made an opening, so we towed and tracked until the ice closed again, and we then made fast. A dark sky was shewing to the southward, and the barometer began to fall, which denoted a strong wind from that quarter. No time was lost preparing for the safety of the ships, and we began to saw docks. The Abram and Lord Gambier were in one, and the Hudson, of Hanover, in another. Our ship was placed in a third, and the Anne brig of Hull, remained sheltered by a point, whilst the Commerce, of Peterhead, was about two miles in the offing, jammed between two loose floes. A furious gale with heavy rain and sleet came upon us.

When the weather cleared, we saw the Commerce a total wreck, and the Anne was badly stove. When the gale abated, every precaution was taken for the safety of the vessels, such as sawing, etc., getting the boats further from the ships and provisioning them, as there were signs of more bad weather coming. The storm came on again and raged with greater fury, and in a short time the Anne also became a total wreck. Momentarily we all expected to share the same fate. Providentially we stood the test, although we had heavy pressures upon us. The Hudson had her stern post started, but the others did not suffer materially. The crews belonging to the two wrecked ships stayed amongst us for a couple of days, then left and made for the Danish settlements with their boats, in hopes of getting home sooner that way, as our condition was so uncertain.

The weather became more settled, but the last gale had broken up the land floe into fragments. We waited for an opening to the northward; instead of that, it appeared in the opposite direction. The south water could be seen from the mast head, so we concluded that to be the best course to take, as the winds seemed inclined to prevail from the S.W.

K

With tracking and towing a few hours we reached the south water, and encountered another gale, which would have annihilated our little fleet if we had remained in the ice. We searched every bight, hoping to get to the westward, until we arrived off Holsteinberg, and were becalmed for two hours near a small bank having about twenty-three fathoms of water upon it. The bank suddenly deepens, with no bottom at seventy-five fathoms.

We tried our luck at cod fishing, but were disappointed of our expected treat. The Esquimaux name for Holsteinberg is Tirieniak Pudlit, or Fox Pit ; this is the principal place for deer upon the east side of Davis's Straits. Upwards of three thousand deer skins are annually sent to Denmark. The hunting grounds are extensive, and the natives go up the fiord for many miles. Salmon is also plentiful there. The rise and fall of the tide is very little compared to the opposite side of the Straits.

A breeze springing up, we reached to the ice edge and saw no prospect of getting within sight of the west land, or into any opening of the ice. We held a consultation, and unanimously agreed to retrace our way back to the northward, though it was very trying to our patience. We saw little change in the appearance of the ice until we came again into Melville Bay. There, a great deal of water was seen, and plenty of room to ply with a northerly wind, but now calms prevailed, which made it very slow, dreary work, towing. After towing, sailing, and taking every advantage of the light winds, we came in sight of Cape York on the north side of the Bay before we were able to strike across Baffin's Bay, and at last made the land on the south cheek of Lancaster Sound.

It was not far from here that the Isabella, whaler of Hull, Captain Humphrey, picked up Captain J. Ross, R.N., and his crew in their boats. They had left their ship, Victoria, fast in the ice in Boothia Straits. The expedition left

England in 1829, and after passing through many privations were rescued in August, 1833. They were supposed to have been dead two years. They had passed the last winter at Fury Beach, named after the ship Fury, which was wrecked there in 1825. At the present time, there is a canister of preserved meat, a stove, and a boat's compass, which were left at Fury Beach by Sir E. Parry in 1825, and brought to England by the Isabella, and presented to the Trinity House, Hull, by Captain Humphrey.

The following is an abstract from Captain J. Ross's voyage in the Victoria:—" Having made arrangements with the John, of Greenock, whaler, to accompany him to Davis's Straits, the men cowardly refused at the last moment, so that Captain Ross had to sail without a companion. The following year, 1830, on board the John, of Greenock, a mutiny took place, attended by death of the master, Coomb, but under circumstances which have not yet been rightly explained, so far as can be understood. The mate, with a boat's crew, was expelled at the same time, and having never since been heard of, are supposed to have perished on the ice. The ship was then put under the command of the Specksoneer, and lost on the Western Coast, when many of the crew were drowned, the remainder being saved by a whaler which was accidentally passing."

We ran to the southward, and when off Scott's Inlet fell in with the other vessels. Most of them had made a good voyage. This was very dishearting to us, as we were so very near to them at the Vrow Islands. If the calm at that place had continued two hours longer, our prospects of making a good voyage would have been much brighter.

Such reverses of fortune are frequent in this country, and shews that perseverance is not always attended with success. One ship may be fortunate in getting whales, whilst another situated a quarter of a mile distant would not be able to get one.

We cruised from Scott's Inlet down to Agnes' Monument and Cape Kater, and many whales were seen. All the vessels had their boats away every day, ours amongst the rest. One day two of the harpooners fired and missed, but the third got fast to a large whale, and we succeeded in killing it before darkness set in, but the wind increased, which made it most dangerous and difficult to flense in the dark, as we were amongst numerous bergs and heavy loose ice. The ship was nearly unmanageable with having a whale alongside. After a weary fifteen hours' flensing, we were able to set our canvas and ply to the northward, having drifted many miles during the night.

The other vessels which had been in our company had the same bad luck, and although we had been amongst so many whales during the season, we did not succeed according to expectation. The crews of the more successful ships heartily sympathised with the less fortunate ones, but it is better to be envied than pitied.

I may here mention that sailing ships had an admirable code of distant signals for their boats. For instance, when a ship is under full sail, and the boats, say two or three miles distant, and a fish appears ahead, the flying jib is hauled down; if astern, the mizzen gaff topsail is hauled down; if on the weather bow, the weather clue of the fore-top-gallant sail is clewed up; the lee bow, the lee foretop-gallant sail is clewed up; on the weather beam, the weather maintop-gallant sail is clewed up; on the lee beam, the lee one is clewed up; to stop pulling, the foretop-gallant sail is clewed up at the masthead and braced by. If this still remains in that position after the boats have stopped pulling, it is the signal to return on board. For near signals, a black waver resembling a frying pan is used by the master in the crow's nest, who points it in the direction he wishes the boats to pull. To return on board, a black canvas · bag, called a bucket, is hoisted at the mizzen topmast head. Those signals are carefully watched. At the present time steamers always keep near to their boats, and the distant signals are not used.

From latitude 69° N., to latitude 74° N., on the east side and Melville Bay, not far from the land, a strange phenomenon is heard, resembling a very weird whistling in a high note, and gradually dying away to a very low one. It is only heard when it is calm, and most distinctly when in a boat or in a ship's lazarette, which is nearly on a level with the water. On the deck it seldom can be heard. When in a boat it is so distinct that one could point the finger to the exact place, apparently, from whence the sound proceeds, and where it dies away. I have often wondered whether it is the electricity of the Aurora Borealis rushing through the air, as it only takes place at night in the summer season. The Aurora Borealis cannot be seen when the sun shines bright at midnight, but can it be heard? I have never known this warning to fail as a forerunner of a S.W. breeze. The louder the sound the fiercer will be the gale. When the sound ceases the dark nimbus clouds begin to rise in the south, and in a few hours it is blowing a gale from that direction.

Captain Parker, during my apprenticeship, and when I was mate with him, never failed to take notice of this warning, and if the ship was not in a sheltered position from the S.W. winds, he would do his utmost to secure a safe place if possible. Many times this prognostication has saved the vessel from great danger.

At the present time steamers have not the opportunity of observing these little incidents, which were of great service to us in sailing ships.

I have before remarked that off Cape Farewell, and in the open of Davis's Straits, in the early part of the season, when the Aurora Borealis has been very vivid, we have invariably had a gale, generally from the S.W., yet I do not say that when it is seen beautifully bright and clear

it foretells a storm. Only when coloured with thick
vapourous streaks. There is not a more splendid sight in
the heavens than a bright clear Aurora, or, as the
Shetlanders term it, the Merry Dancers. When an
echo resounds from a large rugged berg it invariably
foretells a fine north wind. I have also noticed when a
mist issues out of the clefts of the land, especially at Cape
Searle, which is much rented and cracked, it foretells a fog.
Those signs are not noticed now on account of the perfection of meteorological instruments.

But to return to our voyage. Having got the blubber
made off, that is, put into the casks, we remained cruising
about for some time off Home Bay, and saw many fish, but
could not get near them on account of the bay ice, which
had begun to make in the smooth places. We now thought
of getting a supply of fresh water for the passage home, but
could not find a berg or floe with any water upon them.
Our only alternative was to obtain a supply of berg ice,
which is a tedious process, especially with a swell on, as a
boat cannot lie alongside. So the ice has to be chipped
off with lances, and then grappled ou tof the water by hand,
which is not a pleasant occupation when it is freezing keen.
It also gets a coating of salt, which makes it brackish
when melted. After getting some on board it had to be
broken up into small pieces and put into casks.

At the present time all vessels are provided with tanks,
which saves much labour. With this supply I decided to
go to the eastward to see how the ice lay, as it had formed
into a solid body to the southward, and was closing fast in
that direction ; but seeing several fish inshore we reached
in with a fine breeze, and sent the boats away after them.
The wind died away, and the current set us into the bay ;
by morning we were frozen fast, with only one vessel in
sight. We tried to break our way into a lane of water by
milldolling, that is having a boat hung in a tackle from the

bowsprit, level with the water, and breaking the young bay ice ahead of the ship. This precaution of hanging the boat in the tackle is taken for fear a breeze may spring up and cause the ship to run over the boat, endangering the lives of the crew, as I have known to be the case.

After toiling all day, we only succeeded in getting a mile. The s.s. Narwhal came to our relief, and towed us into clear water without the least difficulty. This showed the superiority of steam over sailing vessels. We plied to and fro for some time. The nights were also lengthening, and the weather boisterous with thick showers of sleet and snow. Being now in open water, and no prospect of getting any more whales, we took our boats on deck, and made every-thing secure for the passage home. We worked our way to the southward with a strong S.W. wind. I had no chrono-meter on board to ascertain our position when the weather cleared, which caused me some uneasiness, and we had not seen the land or sun for several days. However, we managed to find our way down the Straits, although the weather continued to be very bad.

With favourable winds at intervals, we reached Lat. 57° N. and Long. 24° W., when we encountered a heavy gale, which forced us to lay to under a close-reefed main topsail three days and a-half, and drifted us about one hundred and fifty miles to the northward. Luckily, we were a little to the southward when the gale commenced, and when it moderated the wind changed to the westward, which enabled us to sail with a flowing sheet.

The first land we sighted was Barra and Rona Islands, then Sumbro' Head, and we soon brought up in Lerwick Harbour. Two ships had arrived the day before, and some came after us. All had experienced the same gale, and been driven to the northward of the Faroe Islands. We waited here three days for a fair wind, and eventually arrived in the Humber, and into dock.

This closes another eventful voyage of varied successes amongst the Davis's Straits whalers, and begins a new era in the whale and seal fisheries, as men having experienced the great difference between steam and sail, few will go hereafter in a sailing ship if they can possibly get into a steamer. My own ideas led me to think that this year would prove the death-blow to sailing vessels. It is so disheartening to men towing and tracking at a snail pace, and a fine powerful steamer pass them at the rate of eight or nine knots. I noticed this year especially that men would cheerfully tow in company with other ships until a steamer appeared in sight, then they lost energy, and the ship would not have steerage way, after seeing the easy life those on board a steamer have compared to their own laborious work.

CHAPTER XV.

FIRST EXPERIENCE IN A STEAM WHALER—SEALING ON THE
NEWFOUNDLAND COAST—HOSPITALITY OF ITS PEOPLE—
BROKEN PROPELLERS—WRECKS ON THE COAST—SUCCESS
FUL WHALING IN POND'S BAY—A PLEASANT VOYAGE.

THE following year—1862—I went chief mate of the
s.s. Polynia, belonging to Dundee, with Capt. Gravill,
senr., one of the kindest men that sailed for Davis's Straits.
He was also a very enterprising and strong-nerved man.
This voyage we were to prosecute the seal fishing on the
Newfoundland Coast, and afterwards proceed to the whaling
grounds in Davis's Straits.

The Polynia was a full-rigged ship, and fitted with an
auxiliary screw capable of driving her nine knots. We
sailed from Dundee on the 10th January, as the sealing on
the east coast of Newfoundland is one month earlier than at
Greenland, and called at Shetland to make up our complement of sixty-six men. A few days after leaving Shetland,
the measles broke out on board. The doctor judiciously
kept the nature of the disease a secret from the
crew, and in about ten days those who had been suffering
were able to return to their duties.

The wind was favourable until we reached Long 42° W.
There we experienced a heavy gale, and lay to under bare
poles for fourteen hours. The ship being heavily laden, we
lay in the trough of the sea for some time before steam
could be got, on account of the seas which she shipped
down the funnel. Then I saw the superiority of steam, for
when the engines got into working order we lay much

easier and shipped less water. We sustained a little damage to the bulwark, and when the gale abated, the wind veered to the N.N.E., and we were able to carry the top gallant sails until we made the ice about one hundred and thirty miles from the land. We sailed and steamed through this ice until we sighted Cape Bacalieu.

When night came on, the ice was so tightly packed that we could not move. This was the first time we had been fast in it with a land right in sight. The following day steam was got up to try and force our way to the land water, if there was any. We got within a few miles of the land, but as there was no water to be seen, our only alternative was to make fast to an iceberg, and wait for the ice to slack, so that we could range about and proceed to the northward in search of the seals. In a few hours a change took place, and we steamed several miles in that direction along the land, and again got tight beset amongst very heavy ice. A strong gale from the eastward began to blow, which drifted us rapidly up Trinity Bay and within a mile of a rock called the White Rock. Then the pressure became so heavy that our position began to look serious. Although our splendid vessel was built as strong as wood and iron could make her, we were obliged to put several blasts of powder under some large pieces of ice which pressed heavily upon our quarter. The cause of this sudden rush was owing to the bay getting tightly packed. It was a grand but awful sight to see the immense pieces running one over the other, then turning upon their edges, falling and crashing like peals of thunder.

In such a case, if anything had happened, it would have been a difficult thing for each man to take care of himself, as it was utterly impossible to launch a boat. The boats were taken upon deck on account of the ice screwing up along the sides level with the rails. But we knew that this could not last much longer, as the ice would soon be solidly fixed at the head of the bay, and could drive no more. We drifted abreast a place called Hearts' Ease, on the north side of the bay. The ice had become quiet, although the wind was still blowing strong.

The following day the weather was calm and clear. The ice presented a strange appearance, being composed of a screwed-up pack, and difficult to travel upon. Four of the men succeeded in walking to the land, and returned with some of the inhabitants belonging to Hearts' Ease, who stated that provisions were difficult to be obtained from the principal places along the coast. The people who came to us were very religious, intelligent, and well educated. After having refreshment, the captain gave them some tobacco and bread, which they greatly needed. They had plenty of money, but provisions were scarce. They said nearly all the sealers were fast in the harbours, and had not been able to get out owing to the winds prevailing from the eastward, and there was such a quantity of ice along the coast. We had not seen our companion, the s.s. Camperdown, since we left Shetland.

There was no change in the ice for two days, when a fine breeze sprang up from the westward, causing it to separate quickly and drift from the land, which enabled all the ships to leave the harbours in search of seals. We steamed to the outside, which now extended but twenty miles from the land in some places, and went as far north as the Funk Islands, where the ice lay a long way to the eastward. Seeing no signs of seals, we prepared to come south.

The next day a heavy swell set in from the eastward, causing the ice to close rapidly. When steaming south we saw several schooners in the ice with their ensigns flying in distress, and some men travelling towards us. We immediately steamed the ship into the pack, as there was no possibility of the men getting to us on account of the outer edge of the ice being in such a broken condition, with the heavy swell. Having steamed half a mile in the pack, both blades broke off the propeller, and the ship lay broadside on to the swell, causing her to roll heavily. All hands were soon at work, and in four hours we had the broken one unshipped and the new one ready for lowering into its place.

During this time twenty-four men came on board, as their vessel had just become a total wreck ; also four other vessels which were further inshore. Their crews had travelled to the land, which was about ten miles distant from them. The heavy swell continued all night, but the next day it subsided. We lowered our propeller and steamed to the outside. The ice had slacked, and the season was getting late, so we steamed towards St. John's, where coals had been deposited for us, and entered the pack a little to the northward of that place, but only again to get beset, and it was so tightly packed that our powerful steamer could not force her way through it. We were rapidly drifting south, and in the afternoon opened the harbour of St. John's, and could distinctly discern the town and Cathedral clock from the crow's nest. It was very annoying to be so near, yet not able to get into the harbour.

At midnight we were off Cape Spear, and knowing that the further south we got the slacker the ice would be, we made ourselves content until daylight, and then made every effort to get into a narrow lane of water, which lay along the land. After some difficulty we succeeded in doing so, about two miles south of a harbour called Bay Bulls, but lost one blade off the propeller. This lane of water only extended half a mile along the coast, towards the harbour, and as the ice was closing rapidly it was a race between us which got there first. Fortunately our one-bladed propeller won the race, although it was a very close one. On entering the harbour the water was not above two ships'

lengths broad between the land and ice. We brought up
not a moment too soon, as it was packed quite full two
hours later. The shipwrecked men left us for their
respective homes, some belonging to St. John's, which was
about eighteen miles away, and the transit was by sledge.

The following day the engineer went to see if there was a
possibility of having a new propeller cast. We were
dubious at first whether it could be done, as the foundry at
St. John's had not cast so large a piece of metal before.
We remained here two weeks, and saw no prospect of
getting to that place while the ice was so closely packed
upon the land. The crew were kept employed in watering
and other work. The blacksmith and mechanics prepared
a new blade for the propeller. It consisted of two pieces
of boiler plate, with the twist or pitch of the blade, and
firmly bolted to the boss. Between the two pieces
of plates it was solidly filled with wood and bolted tight
together. We were ready whenever a change took place.
Some of the officers went to the top of the hill morning
and evening with a spy-glass to look at the state of the ice
outside the harbour.

One evening they observed a vessel in flames, and three
more apparently in distress among the ice about six miles
away, and a number of men travelling towards the land.
Our men at once returned on board. The boats were
quickly manned and pulled to the ice edge, and rescued
about sixty men. It was a fortunate thing that they were
seen before dark, or they must have perished, as the state
of the ice would not have permitted them to reach the
land. They were supplied with hot coffee and a hearty
meal. Our crew gave up their berths for them that they
might get the rest which they so much needed.

In the morning, after breakfast, they went on shore to
their homes, some having many miles to travel. This
place contained 300 inhabitants, who were most kind

and hospitable. I shall always think of them with
pleasure. Their treatment was so different to what had
been represented to us. Their kindness was the more
emarkable as our steamers were likely to supersede their
sailing vessels in sealing. A few years after this the Dundee
fleet of steam whalers made successful voyages to Newfoundland.

At last the ice slacked. We weighed anchor and steamed
towards St. John's. Night overtook us before we reached
that place, and the ice being closely packed outside the
Narrows, we struck a heavy piece with the propeller, which
broke off the iron blade and crippled us again. However
we managed to get safely brought up in the harbour.

The following morning we got under weigh, and moored
alongside the depot, on the south side, and commenced
coaling. We sent our broken propellers on shore to be
fitted with blades similar to the one we made in Bay Bulls.

We were detained here four weeks, sometimes laid in the
stream and sometimes at the Galway Wharf. The harbour
of St. John's is land-locked, and well protected, and it can
accommodate a large fleet of shipping. The entrance is
called the Narrows, and is a miniature Gibraltar with its
fortifications. The island is 420 miles long, and 300 broad,
but the interior has not been explored. It is said that
so far as is known it is very fertile, and abounds in deer,
bears, wolves, etc., also some large and magnificent lakes
abounding with salmon.

When we were staying in Bay Bulls we only went about
three miles inland, and saw plenty of small spruce fir-
trees. We cut down some of the largest for boat hooks, etc.
Several schooners came into the harbour of St. John's with
seals, but none had great catches. They had got them
during the time they were beset south of White Bay, and
reported that thousands of young seals were killed by the
ice running over them when it pressed on the land. It

appears that the greatest quantity of seals were to the north-ward this year, and a large number were killed by the inhabitants.

This was a most disastrous year. About fifty sail were lost on the ice along the coast during our stay, and it was the most severe time experienced by the oldest sealer in St. John's since 1831. Our vessel being the first steam whaler that had been in the harbour, great interest was taken in her. Every fine day numbers of visitors came on board to inspect her, and the whaling gear.

The Governor, Sir Alexander Bannerman, and his officers, visited us, and gave us great credit for having the ship so neat and clean. In course of conversation, it was found that Sir Alexander and Captain Gravill were old acquaint-ances, as they had met in Greenland when boys. The newspapers also spoke well of the conduct of our crew, and wished us every success in whaling, although they said they were not sorry we had not succeeded in sealing upon their coast.

At last a new propeller was cast and the other one repaired, which we shipped, and were once more ready to seek our fortunes in the far north. It was with regret that we parted with our kind friends in St John's. Staying here would not bring grist to the mill, so we bade adieu to them, and steamed away towards Davis's Straits. We sailed and steamed along the edge and sometimes among streams of ice, expecting to fall in with the bladder-nose or hooded seal, as good voyages have frequently been picked up after the other sealing was finished, but our expectations were not realized. In Lat. 55° N., the ice led us further to the east-ward, so we made the best of our way to the whaling grounds.

I will here remark that the s.s. Camperdown, of Dundee, came into the harbour three days before we sailed. She had been beset six weeks off White Bay, and had also lost a propeller. She took her coals and proceeded on her way.

Off Black Hook we spoke the s.s. Narwhal, of Dundee, which had brought us a strong, new propeller. We were now well provided, having two spare ones. We steamed in company with her, and reaching Melville Bay, the weather being calm, we made fast to a large floe, during a dense fog. When it cleared up we steamed into the Bay among large floes, occasionally ramming at necks of ice which barred our progress. Then we came in sight of Cape York, and got into the open north water off Cape Dudley Digges. We were only twenty-six hours getting through the Bay, whereas a sailing ship would have thought it a quick passage in fourteen days. Not an hour's loss of rest was experienced by the crew, but the captain was at the mast-head most of the time. This shows what an easy time sailors have on board a steamer, compared with those in a sailing vessel. We lost no time in steaming to Pond's Bay, and sighted the land on the north side. Many large floes lay in our path before we came to the land floe, which extended about twelve miles. There was a favourable bight in the open of the Bay, and we steamed in that direction, but a large floe impeded our progress. We charged the opening at full speed. To all appearance it was only two and a half feet thick, but judge our surprise when we struck it. The shock made the pitch fly out of the seams of the deck, those who were on deck were thrown down by the concussion, and the ship trembled and rolled as though she had been at sea; we found it about five feet thick instead of two and a half.

I have often thought it strange, when I have read of Atlantic liners running into large icebergs at such great speed, that they do not receive more damage. Our speed was not more than seven knots per hour, and this was only a floe which split at our bows, but an iceberg would not have moved an inch. However, we steamed to the north part of the bight, and making fast, sent two boats on the watch, *i.e.*, a boat fully manned and equipped, lay at the floe edge ready to fasten to a whale. We had not been long waiting before we struck one, which was soon killed ; by the next day at noon we had killed four. The Narwhal was also successful. During the time we were putting the blubber away, the other Dundee steamers hove in sight.

The Camperdown seeing our position so favourable, sent his boats too near ours, which made both parties liable to frighten whales which might rise between us. This was contrary to the usual practice, and gave great dissatisfaction to us. Formerly captains were more just in their dealings, and were always ready to give a helping hand when required. Whenever a harpooner from another ship gave assistance, either in lancing or striking a friendly harpoon, it was the custom to present him with a pair of canvas trousers and a pound of tobacco, and a pound to each of the boat's crew. We steamed our ship about a mile further from the Camperdown, and during the time we were moving he fastened to his first whale. Eight days after this occurrence we had captured thirteen.

To shew how Dame Fortune distributes her favours. We were a company of eight steamers, and lay a certain distance from each other at the floe edge. Each vessel had two boats on the watch. The s.s. Tay was in the middle and apparently in the best place. We captured from nine to twelve whales each, except the Tay, and he only killed one. Several rose close to his boats ; the harpooner several times had his hand on the trigger line ready for the fish to rise again, but only to be disappointed. The same fish would rise close to the bow of another vessel's boat, so that they had no trouble to pull, only to fasten to it sometimes with both gun and hand harpoons. To my knowledge this occurred a dozen times. In such cases as this, no blame can be attached to the captain or men.

L

This year the bomb lances were frequently used by us, with astonishing effect. On one occasion a whale was going under a patch of ice, and no doubt we should have lost her, had not the harpooner fired a bomb lance, which exploded in a vital part that she rolled over quite dead.

I will give an instance of the strength of the whale. As we were killing one, which lay alongside the floe, so that only one boat could lance, I and another went on the ice to assist. In her dying agony she struck the ice, which was three feet thick, with the flat of her tail, breaking it in pieces, and making the floe shake the length of a ship from her.

In some cases, a fish could not be approached with the hand lance when they were lively, hence the benefit of the bomb-lance. Great precaution must be used with them, the powder in the gun must not be pressed too tight, or the fuse is liable to bend, and in consequence it will not ignite. Another time one exploded at the muzzle of the gun, and rent the barrel down to the breach like a piece of paper, some fragments struck my boat's bow, and several pieces entered the wood, but fortunately no one was hurt. It was a foolish action to fire when two boats were in such close proximity. I was on one side lancing the whale, when the boat on the opposite side fired.

The loose ice now closed upon us, so that we could not ply the boats, and the whales were very numerous. At every hole one was to be seen, and it was annoying that we could not get at them. The following day, the ice slacked off, but they had all disappeared. After waiting several days, we went further south and killed two more, which made us fifteen. Up to this time fortune had crowned our endeavours, but now deserted us. Many whales were seen but we could not get near to them.

This season every whale that was fired at, we succeeded

in killing, so different to the previous year, when so many were missed, thus shewing the difference in the ability of the harpooners.

We were now at the north side of Home Bay, the ice was closing rapidly, and a fog bank travelling quickly to the westward. We therefore made fast to a small iceberg, aground close to the land, and shot two bears, which brought our number up to eight. A boat's crew, with three rifles, went on shore, and returned with two large deer.

Captain Gravill gave the largest to the crew, which made them a delicious meal. Whatever is brought on board by any of the crew must be delivered up to the captain. It is very pleasant when a ship's company work in unity, and have a commander whom they respect. Such was the case with us, and I must say that this was the happiest voyage I had spent since I had been to sea. The sanctity of the Sabbath was always kept. Divine Service was held in the cabin by the master, and everyone attended who was not on duty. It prepares a man for the toils of another week, and seems to bring him nearer to his friends at home. No boat was allowed to be lowered on the Sabbath day, unless it was absolutely necessary. Eventually we came to Cape Hooper, and brought up in the harbour.

There were two sailing vessels at anchor in this place, which had not seen a whale during the season. They only stayed two days after our arrival, and went further south. For some time we were alone, and daily sent our boats away to the outside.

One morning when we were away in the boats we gave chase after two fine deer that were swimming from the low land on the north side of the harbour to the island on the south side, from which Cape Hooper derives its name. It was astonishing to see how fast they could swim. Nevertheless, they reached the land before we were within gunshot of them, and soon shewed us a clean pair of heels. This day proved very boisterous, and we dare not venture far from the land, so therefore kept under the lee of the island. A swell on the rocks prevented us from landing to keep a look-out.

As evening drew to a close, the signal was given for the boats to return to the ship. We were then about five miles away, and the wind was blowing a gale out of the harbour. After pulling four hours, we came within a quarter of a mile from the vessel, but could not get any nearer, being too exhausted and drenched with the spray. Buoys were streamed from the ship with lines attached to them, but they were ineffectual, so we had to abandon all hope of being on board that night, and returned under the lee of the island until daybreak.

It was a most miserable night. All were wet through, and as we could not land, we rode with our grapnels as close as possible to the rocks and laid in a row, a boat's length apart. Each rigged their mast with a sail to form a tent to shelter them from the pitiless storm. The snow fell heavily, and only half the boat's crew could be sheltered by the small tent or sail. Never did a night appear to be of so long duration. We did all we could to cheer each other, and, notwithstanding our serious predicament, we jested upon the comfortable time we were having.

In the darkest hour an incident occurred which threw us into a state of commotion. A loud roar warned us that a bear was amongst the boats. We could not see Bruin, but had our rifles ready to receive him. In a little while we heard another roar on the rock close by. Snap went the guns in that direction. Whether we had wounded or frightened the brute could not be ascertained, but we heard him no more.

This occurrence kept us on the alert all night, and everybody was glad when daylight appeared. The weather had moderated, and we did not go on board, but followed the usual routine of rock-nosing until night, then pulled to the ship, and were heartily glad to change our clothes and get to bed. After leaving our place of shelter, we found out that if we had pulled to the south part of the island we could have landed on a fine sandy beach and smooth water, where we might have rigged our tents, and spent a more comfortable night on the land, but we thought the wind was more on that side.

On another occasion, the ice set into the harbour and we could not get the boats within three miles of the ship. There was no place to beach them, on account of the rocky nature of the land, and it was not safe to leave them on the ice, so the only alternative was to launch them to the ship, but the ice was in such a broken condition that it took two crews to launch one boat. By midnight we had three boats on board. Many men had slipped in the water, and all were more or less wet.

After staying on board a short time we returned for the other three boats, and by six a.m. had them also safely at the ship. The watch was set, and everybody got thoroughly refreshed. During the whole of the day it snowed heavily.

The following day the wind came from the westward, which cleared the loose ice out of the harbour, so that the boats could go away again. While they were away the wind freshened, and brought down a large piece of floe which had broken from the head of the harbour, and was drifting in the direction of the ship. Those on board pulled to the floe and drilled holes in it before it reached us, and put in several powerful blasts of powder which split it in several pieces. If we had not taken this precaution, probably the chain would have parted, and we should have lost an anchor and seventy-five fathoms of chain cable before steam could be raised. As it was, the strain upon the windlass was very heavy.

Another day, as we were manning the boats, the snow began to fall. The captain was not a man to send us away unnecessarily, so he gave orders for us to remain on board. The ship was in perfect safety, and ready for sea.

We requested him to let us have a run upon the island in search of the two deer I previously alluded to. He granted us permission, and most of us were soon off like schoolboys on a holiday.

The island lay east and west, and about four and half miles long and one broad. We spread ourselves in a line, and travelled from one end to the other. Although the snow was falling fast, we thoroughly enjoyed the change. We saw the two deer, but they sighted us first, and off they went up a very rugged part of the island and left no trace behind them. At the east end we shot three hares and thirty-four ptarmigan, and chased several ermine, but they too escaped among the rocks. We returned on board after four hour's recreation, and set the watch.

The next day the s.s. Narwhal came into the harbour, and caught the deer as they were swimming across. When they were dressed, they weighed eighty-four pounds each, without their heads. The captain kindly sent us half of one. We shot several blue and white foxes on the north side. We also killed a large walrus, which we hauled upon the beach and flensed.

The next day it was partly devoured by wolves, as there were many footprints seen in the snow. I had formerly heard from a family of natives, who used to make this place their winter quarters, that the wolves became too numerous, and they were obliged to leave. They said their dogs would sooner encounter bears than wolves. I have not been within gunshot of any; those I have seen so much resemble a large Esquimaux dog, that at a distance it is impossible to distinguish one from the other.

At the head of this harbour there is some beautiful

scenery, but no civilised person has been there. I had a splendid view from the summit of a mountain on which I climbed to ascertain the state of the ice to the northward. Inland there is a large sheet of water extending far to the southward, but I could not say whether there is any outlet, as it has not been explored. In many parts of this country what is supposed to be the mainland often turns out to be a large island. The way up the mountain is most difficult and dangerous—very rugged, and has all the appearance of volcanic origin. It appeared to me that the S.E. side had burst open, and thrown the fragments of rock on either side.

The s.s. Narwhal steamed out, and left us alone. We remained until the young ice prevented us from pulling the boats, and a great quantity of ice drifting along the land, we weighed anchor and steamed outside. During the last few days many whales were seen, but it was impossible to get near them.

We were the last vessel north. Some had gone to Cumberland Gulf. Once clear of the ice, we set our sails to a fresh northerly wind, homeward bound. In due time we arrived safely in Lerwick, discharged our men, and under steam and canvas soon sighted the Bell Rock, and entered the river Tay. We were not long in getting into the dock, and we began to discharge the next day. Thus ended a most pleasant voyage.

CHAPTER XVI.

IN 1863, I was once more mate of the barque Emma,
now sold to Messrs. Gilroy, of Dundee, in command
of Captain J. Nichol. We shipped our crew of fifty-two men
in this port. A dense fog set in on leaving. We brought
up off Broughty Ferry for two days. The wind changed to
the southward, and with a fine breeze we weighed anchor
and proceeded to sea.

We had not got many miles away when the wind backed
to the eastward, and blew strong with thick blinding sleet
and rain. As we were near the land, and on a lee shore, we
were obliged to carry a press of canvas to make her weather
Kinnard's Head, which we accomplished, owing to her being
a fine, handy ship.

When we arrived in the Pentland Firth the weather
cleared, and we soon sailed through it, passing Cape Wrath,
the Butt of Lewis (this night—10th March, 1863—
was the first time a light was shown on this point, in com-
memoration of the marriage of the Prince of Wales), and the
island of St. Kilda, and began to prepare for the voyage.

So many things have to be made and repaired that it
requires the crew to be constantly employed during the
whole passage out. Everything went on prosperously until
one very dark night, when the watch was furling the main-
sail a man fell overboard from the mainyard and was

drowned. The wind blew strong, and the sea was running
high ; but, owing to the darkness and storm, he was never
seen again. If it had been daylight he could not have been
saved ; and, strange to say, I do not think more than six
men out of the whole number could swim. This was the
poor young man's first voyage to sea. It was not our practice to allow an inexperienced hand to go aloft until we
arrived in smooth water. He had been teased about
staying on deck by some of his thoughtless companions, so
he ventured aloft for his first and last time.

On reaching the longitude of Cape Farewell, strong gales
from the north-west prevailed, and we were under close
reefed, and at intervals, double reefed topsails, for three
weeks. It was a long, cold, dreary time. We were constantly in sight of bergs, which kept us always on the alert.
When a fair wind came, we found ourselves in the same
latitude and longitude we were in three weeks before.
However, we struggled on, and eventually made the ice off
Baal's River, on the east side of Davis's Straits.

This place is frequently spoken of by Baffin, in his first
recorded voyage in 1612, in which Mr. Andrew Barker was
master of the second ship called the Heart's Ease, belonging
to Hull.   Mr. Barker was admitted a Younger Brother of
the Trinity House, Hull, in the year 1594, and was warden
in 1606, 1613, 1618, and there still hangs in the Hall of the
Trinity House, a kayak or canoe, with the effigy of an
Esquimaux. It has the following inscription, " Andrew
Barker, one of the Wardens of this House, on his voyage,
anno domini 1613, took up this boat and a man in it, of
which this is the effigy."   It is still in good preservation,
and has the original paddle, drogue, bird dart, harpoon, and
gear.

There is another canoe in the Schiffer Gessellschaft, at
Lubeck, brought to Europe by the Danish Expedition, in
1607. At this early period the natives were savages, but

after the Danes settled amongst them, they gradually became civilized.

When we came to the fishing grounds off Disco, we killed a whale, which was the smallest I had seen upon the east side, with the exception of those which had not left their mothers.   This season the ice was more tightly packed upon the east coast than had been seen for several years. Is was with great difficulty we reached N.E. Bay.   A few whales were captured amongst us, and when the fishing in that place was over, we directed our course northward in the usual manner, until we came to the Duck Islands, and nothing but large floes presented themselves in Melville Bay.

We waited in that neighbourhood some time, moving about from place to place according to the state of the ice, and seeing no prospect of getting through, we went south. The steamers had long since left us, and were no doubt reaping their harvest in Lancaster Sound or Ponds Bay, while we were retreating.   I think that sailing vessels in this country were getting more accustomed to retreating than advancing.   We had a succession of fogs and adverse winds   All our officers had been used to steam of late, which made us sometimes forget that we were in a sailing ship.

From the time of leaving the Duck Islands, until arriving on the west side, we were continually on the move, regardless of the weather.   The ice formed a compact body off Cape Broughton.   Neither whales nor anything else were to be seen, in fact it appeared as though life had deserted this part.   There was very little water between the east and west ice.   The frequent fogs caused us much anxiety, and kept us always on the alert.   We almost gave up all hope of success, yet there were the harbours to the south still open, where many good whales had often been caught.   The ice kept driving us further and further south, the nights were getting longer and darker, numerous bergs were in the vicinity, and showers of snow were more frequent.

One night when off Cape Dyer, and among a number of bergs, a very heavy swell set in from the southward, accompanied by thick snow. All hands were called to hold themselves in readiness in case of emergency. There was such a light breeze, that the ship had not steerage way. The noise of the sea breaking upon the bergs sounded weird and dreadful, and the blinding snow prevented us seeing how far we were from them, and being in the centre of this cluster, made us fervently wish for steam, and earnestly pray for daylight, which came at last, and with it a breeze that took us clear of the surrounding dangers.

The snow ceased falling, and we directed our course towards Cumberland Gulf, and in a few days brought up in Niatlik, where we found the Alibi and Sophia, both of Aberdeen, and a small schooner called the Franklin, belonging to New London, U.S., commanded by Captain Buddington. Those three vessels intended wintering in this place, and had pledged themselves to assist each other, the Alibi and Sophia on the one side, and the American schooner on the other. This pledge was broken by the former party shortly after our arrival. One day, three boats belonging to our ship followed a whale several miles, which led us into a small bay some distance from the harbour. We were engaged with her about four hours, and were several times nearly within reach of her, but she contrived to escape, and made for the offing. One of the American boats manned by Esquimaux, was pulling on board, when the fish rose close to his boat and he struck her, we went to assist him, as no other boats were near, except one belonging to the Alibi. When coming in sight of the ships, they saw us and sent more boats. Already one of our harpooners had fired his harpoon into the fish, and I gave her several lances. By this time the other boats were with us, and she was quickly despatched and taken in tow. In justice, this whale belonged to the American, but judge our surprise next morning to see the Alibi with it alongside, and the crew beginning to flense. On passing the American, I asked the reason why the Alibi had taken the whale, and he said they had claimed the half for assisting. This was contrary to their contract. If there was any right of claiming, we deserved it more than they, as we had given the most assistance, but such a thing was far from our thoughts. In this case it was might against right, and was a most unjust action, shewing their selfishness, as they took more than half. The boat belonging to the Alibi merely lay alongside of the American's boat in case they required more line. However, fortune favoured the little American, as he filled his vessel before the winter set in without any more help. The other two, after wintering, arrived home the following year with very little.

One evening I paid a visit to Captain Buddington, and during our conversation an Esquimaux lad came into the cabin. Captain Buddington asked him to tell us about Frobisher and all his proceedings in the Straits which are named after him. It was very interesting to hear the lad talk about things which happened nearly three hundred years ago. Captain Buddington could speak the language (if it can be called one), like a native. I could understand most of it, and many things he told us corresponded with what I had read in Luke Fox's voyages. We asked him how he knew these things, and he said his father and mother told him when he was a little boy, and in that way traditions are handed down from generation to generation.

Most of the natives who were here belonged to Nugumut. Some were partly dressed in European clothes, and they could drink rum and swear round oaths in English. Those ships intending to stay the winter only carry a limited number of men from home, and make up the deficiency with Esquimaux, whom they feed during the winter, and give them presents in return for their help.

We saw many whales in this part, but had not the good fortune to kill any. A Scotch brig came from Kemisuack, and joined us here. The young ice was rapidly making in the small bays, and it froze so keenly that our oars had frequently to be taken in and the ice broken off them. Early one morning we got under weigh, and sailed outside of the islands which formed the harbour. A number of whales were in the offing. Six boats were lowered, and sent in pursuit. We pulled for four hours without success.

Then a breeze sprang up, and caused a short nasty sea to rise, which made it difficult to manoeuvre the boats. As night came on we were obliged to go on board, very much disappointed. We dodged about during the night. At daylight we found ourselves some miles from the place, so we bore up for home. When clear of the land, we perceived an ice blink to the northward ; by that we knew the ice was coming down in a body, but so long as the wind continued northerly the Gulf would remain clear of heavy ice. Yet a great barrier might form outside and make it difficult to get out. This afterwards proved to be the case.

The day after our departure, the s.s. Camperdown steamed into Niatlik, and in the course of a few days killed several large whales. The winter had set in, but it was mild, and the fish came near to the harbour, and many close to the ships, so that the boats had no distance to pull, and immediately a boat fastened to one. The steamer buoyed his cable and steamed alongside the boats, and when the fish was killed it was soon towed into the harbour. He got five large whales in three days with very little trouble. A sailing ship could never have done this, as they are not able to follow their boats as a steamer can. The master of the Scotch brig, which came in before we left, was advised to leave immediately by the captain of the s.s. Camperdown, who informed him of the heavy body of ice drifting past the entrance of the Gulf, which would probably cause it to freeze over very soon.   He ignored this advice, and trusted to the steamer towing him clear.   However, he weighed his anchor the day before the s.s. Camperdown. When about twenty miles from Niatlik, he was frozen up and unable to move, although he had a fresh breeze and all sail set.   He pulled his ensign union down for assistance when the steamer passed.   The captain asked what he required, and he answered that he wished to be towed clear of the Gulf.   The captain of the steamer replied that it was an impossibility for him to accede to his request, as probably he would not be able to get his own ship through the ice outside, so he advised him to retrace his way back.

Eight months afterwards, it was reported that the brig got safely back into the harbour.   The wintering ships helped him with provisions, although he had a good stock on board.   If that had not been the case it would have told heavily upon them all.   When the s.s. Camperdown arrived at the outside of the Gulf, it took her three days to get through the ice.   The keen frost had cemented it together, and nothing would have saved them had not the ship been strong and powerful.   This is another illustration of the superiority of steam.   Steamers, it is true, run great risks, but, taking everything into consideration, they do not encounter so much danger as a sailing ship.   It was fortunate we took time by the forelock, or else we too might have been in the same predicament as the brig.

After a fine passage, we passed the island of St. Kilda, and the wind prevailing from the southward, we took the advantage of keeping nearer the land in smoother water, until we neared Cape Wrath, when it became calm, with a heavy swell from the northward, which drove us into the Minch, and in close proximity to a very steep island.   No soundings could be found at sixty fathoms, and the swell was so heavy that nothing could save us but a breeze, or getting the ship's head to the westward. A boat was therefore manned, and an attempt made to tow the ship's head round. The night was very dark, and the ship rolled heavily, which made it a difficult task. A light breeze sprang up, and the sails being trimmed to it, we gradually edged from the land and into the Pentland Firth. A boat came from the island of Swona, and the crew told us that about the time we were becalmed a small coasting schooner drifted on the rocks and became a total wreck. In the Firth the tide runs with great velocity, and with a spring flood and a westerly wind it attains the rate of eight or nine knots. We got safely through into the North Sea, and were soon in the river Tay, and moored in the dock.

CHAPTER XVII.

IN the year 1864 I went mate of the s.s. Polynia, with Captain Gravill. This time we were to prosecute the seal fishery in Greenland, and, if not successful, to proceed as far north as possible in search of whales. We were well manned, provisioned, coaled, and in every way fully equipped for that expedition. Not a single thing was wanting. Our orders were that if we got seven or eight thousand seals to return to Dundee, discharge, re-coal, and again proceed to Greenland. On leaving Shetland, we continued our course until we sighted Jan Mayen's Island, and cruised among the ice in search of seals. The sealing grounds range from lat. 68° N. to lat. 74° N. and long. 2° W. to 16° W.

About the 24th March is the time for them to have young ones. Sometimes they are in one large body ; at other times in two. Late one afternoon we came to a large stream of ice, on which were a quantity of seals with their young ones, about a day old. All hands were soon on the ice. The old seals took the water when they saw us coming. We had twenty-four Minnie rifles, and the others had clubs. The young seals were quite helpless. At first we refrained from killing them, but shot the old ones as they made their appearance in the water. If they did not soon shew themselves, a man would nip the young ones to make them cry, the mother would then appear and was shot. Then we killed the young ones as they would have died of starvation.

Of late years there has been a close time for seals, catchers not being allowed to kill them until they are nine days old. The cross shooting was very dangerous with such long ranged rifles, as a seal must be shot through the brain or it will sink and be seen no more. The bullet often swerved at an angle of 45°, and came with its unwelcome sound of ping-ping amongst the men who were scattered in all directions, and it was a wonder no one was shot.

When evening was closing the signal was given to return on board, but the ice suddenly opened and left a number of men on different pieces, from which they could not jump from one to the other, and it was two hours before all were safe on board. We had shot and killed upwards of two hundred seals, and we knew that the main body of them could not be far away to the westward.

During the night the wind rose from the east, causing a heavy sea. The gale continued three days, most of which time it was snowing and freezing keen. We steamed through streams of ice and dodged in holes of water, shifting our position as the ice closed or opened out. The sky cleared, and when the gale abated we found ourselves alone a little to the southward of Jan Mayen's Island.

Before proceeding any further I will give a short description of the island :—Jan Mayen's Island was discovered by a Dutchman of that name in the year 1611. It lies lonely in the middle of the wide, deep sea between Norway and Greenland, Iceland and Spitzbergen ; and is distant about two hundred and fifty geographical miles from the coast of Greenland. It is twenty-nine miles in length, and six in breadth, and so thoroughly mountainous, mostly with rugged rocks reaching down to the sea, that it has really only at two parts a flat beach and so-called landing-place. The north-east part rises to a height of nearly 7,000 feet in the lofty Beerenberg, which has a large crater, and indeed the whole island is of volcanic origin. In the year

M

1732, Burgomaster Anderson, of Hamburg, reported a
decided eruption from a small side crater, which had been
reported by a seaman ; and in the year 1818, Scoresby and
another captain saw great pillars of smoke rising from the
same place.

Surrounded by floating ice the whole winter through,
and often for a longer period, Jan Mayen lies in the spring-
time and early summer so near the edge of the ice-pack,
that from 1612 to 1640, it afforded to the English and
Dutch whalers a comfortable and much sought after station
for their booty and oil preparation. It is said that a single
ship in one year then brought home from Jan Mayen
196,000 gallons of oil. Wishing to make an attempt at
colonizing, in 1633-34, seven Dutch sailors passed the
winter here.

The small community outlived the severity of the winter
without much danger to their lives, until the scurvy broke
out amongst them ; and as they could not procure the
necessary fresh nourishment, the sickness made rapid
strides. The first died on the 16th April, and all the others
shared the same fate one month later. Their diary ended
with the 30th. When, on the 4th of June, the Dutch fleet
appeared off the island, they were all found dead in their
huts.

Scoresby visited the island in August, 1817, and gave the
first reliable account of it. The interest excited by this
account led to two other visits. Lord Dufferin landed on
the north side of Jan Mayen in 1856, but what with fog and
floating ice, could only stay one hour. In 1861, Dr. Berna
of Frankfort also landed in two places, and the weather
allowed him two days' investigation.

To return to the voyage. We sailed to the S.W. among
much young ice, called by seamen pancake ice, on account
of its first forming in round patches. Our captain's
experience led him to believe that the last gale had drifted

the ice in that direction, with the main body of seals upon it. We cruised about a day or two but found no trace of them. We were boring into the heavy pack under canvas and going about two knots an hour, and passed pieces of ice with numbers of young seals upon them. Our latitude at noon showed us to be in Lat. 69° 10' N., one hundred and ten miles southward of the island, yet it was distinctly seen from the deck.

The seals were now in their best condition, their mothers having left them to provide for themselves ; we knew that any day they might leave the ice and migrate further north, which they usually do when they are three weeks old. On account of it being Sunday, the captain very reluctantly gave his consent to us killing them.

In a few minutes everybody was on the ice with his club, knife, steel, and line, called a "lowery tow," which is used for dragging the seals on board. By 8 p.m. we had killed 1,800 seals. The ice was now tightly packed, but there was a long heavy swell. All hands retired to rest, with the exception of three men, who were on watch to give timely warning in case of necessity.

Although the heavy swell caused the ship to thump on the ice, it did not disturb our slumber. At daybreak all hands were called to get breakfast, and then commenced another day of slaughter.

Eight or ten men, who were not accustomed to the work, were unable to get out of their beds, on account of the sinews of their limbs being contracted. The first day or two affects everyone more or less.

Sealing is very laborious work, especially to those who are not familiar to it. When there is much snow on the ice, and a swell, it is very dangerous, and great care must be taken when a man jumps from one piece to another, for, if he should fall between them, he would be crushed to death in a moment. The weather was extremely cold, which

answered our purpose, as the exercise kept us very
warm.

We were among the seals four and a half days, and killed
an average number of 2,500 per day, although at times we
had not more than thirty men from the ship, the remainder
being on board making preparations for stowing the skins
and blubber away. Our coal had been husbanded very carefully, in case we did not get among the seals, knowing
we should require it if we were not successful. Our tanks
and bunkers were full of coal, so we had only room in the
between decks and top of the tanks. Every space was now
filled up with seals below and on deck. The ship was very
deep, and the swell caused her to roll, which made it
difficult to work. We flensed the skins, *i.e.*, to separate the
fat from them. This part of the business is given to
experienced hands, because if the skin is cut, it is not so
valuable. Some were employed salting them ; others getting
the coal out of the tanks, and some others were putting the
blubber in as the tanks were emptied. Each tank held
five tons, and most had to be got out by hand until so
far down that a shovel could be used. This work took
three days.

Sometimes a few men were sent away to kill seals whenever the ship drove through a small number. If we had not
had so much coal, we could have got twenty thousand ; as it
was, we were obliged to be content with about fourteen
thousand. We worked hard to make room for more, but by
that time the seals, which had escaped the club or knife,
had taken to the water, and disappeared.

The ship was near to the outside, therefore we steamed
towards home with a head wind. During the whole passage
all were employed flensing skins, and putting the blubber
away ; then commenced scrubbing and cleaning the ship.
In a short time no one would have imagined she was laden
with seals. No dirt or grease was to be seen. When all

was accomplished, we sighted Lambaness or Unst, the most northerly part of Shetland. The whole time of sealing, the cold was most intense with showers of very light, drifting snow, which froze on us as it fell. Our beards were hung with icicles of blood, our clothes were bespattered with the same, and we scarcely knew each other when we passed. Seals are so full-blooded that when an artery is cut it spouts a long distance, and very often in the face.

Those which we caught were called " white coats," or the young "saddle back " or " harps." They are covered with a beautiful white fur about an inch long. When their mothers leave them, the fur gradually comes off, and the hair shows itself on the body. I must confess that sealing is very cruel work, but the fiat of a remorseless and callous fashion demands the sacrifice, and the demand must be met by the supply. The poor animals have a most beautiful large, round eye, especially the young.

When a sailor is killing one, it looks up pitifully and cries "mamma " quite plainly, but the club or knife descends, and in five minutes its skin is off. Many times they recover consciousness after being flayed, and will crawl away a couple of yards until the body is frozen stiff. They are very tenacious of life, and, unless the brain is crushed, they will live a long time after the knife has been plunged into the body. Perhaps a more humane method might be adopted in killing them, but at the time of which I am writing we used the best means furnished for the purpose.

The bladder-nose or hooded seal, and the larger species, are not so easily despatched, as they will show fight to a boat's crew when attacked, so guns are generally used. We were eight days from the sealing grounds to Shetland, having steamed the whole way with a head wind, but not much sea. On reaching Lerwick we discharged the men without coming to anchor ; then steamed away for Dundee, and brought up off the town. The s.s. Narwhal soon

followed with 3,500 seals. At tide time we entered the
dock, and the following day began to discharge our cargo,
which took us nearly three weeks. Having taken in coal
and other necessaries, we once more started for the northern
part of Greenland upon a whaling expedition.

We called at Lerwick and shipped the same men, and
once more fell in with the ice north of Jan Mayen's
Island ; then steamed to the westward among very heavy
pieces towards the Liverpool coast, and came to the land
floe, which extended about twenty or twenty-five miles from
the land. This ice was very heavy indeed, and appeared to
be several years old. There were a few small icebergs frozen
in the floe far inshore.

The scarcity of bergs in Greenland plainly shows that
there are not many large glaciers. When sealing we saw
none.

We steamed along the floe edge when it was possible, and
once sighted the Pendalum Islands. The floe, and large
pieces which had broken from it, were nearly of a uniform
thickness of twenty-one feet. The snow upon it was about
four feet thick and very solid ; therefore, when we made fast,
we had to carry the ice anchors nearly a warp's length from
the ship, whereas if the surface had been solid ice instead of
snow, one-third of that distance would have sufficed.

We had to be careful how we touched this thick ice with
the ship, as it would not in the least give way to a blow,
and had also to be very watchful lest a sconce piece should
drive on us, especially in foggy weather, which was the case
six days out of seven. Although at times the sun shone
brightly overhead in a clear sky, the constant fog was very
discouraging, as the ice was so favourably situated for
whaling.

Off the island of Bontekoe, which is a very good
fishing ground, the floe only extended seven miles ; a party
proposed to travel to it one day when the weather was

clear. The ice was as level as a bowling green, not a ridge or hummock to be seen, but a dense fog set in and spoilt our intended walk. The whole time we were here, only one whale was seen, and it was going northward like a racehorse. This was a most notable year, as the whole place was nearly void of life. Narwhals, seals, and birds were very scarce.

We shot twenty-two bears and brought two home alive, which were exhibited in the Greenland Yard, at Dundee, for the benefit of the Infirmary. The bears were very cunning, and we had to be cautious, as most of them were shot upon the ice.

On one occasion, I had severely wounded one through the neck, being a long distance from it ; but it escaped, as it could run faster than us. We watched it from the masthead for three hours, then it returned towards the ship. Four men went to meet it, and one got close to and pulled the trigger of his rifle, but it would not go off, owing to it being at half-cock, which in the hurry he had overlooked. The animal was about to seize him when one of the harpooners shot it through the head.

Another time one came leisurely towards the ship, when we were made fast to the floe, and three of us formed ourselves in a triangle and walked towards the animal. When about a ship's length off it stopped, and we all fired at the same time. Feeling sure we had killed it, imagine our surprise to see it walk towards the man at the apex of the angle. There was not time to reload properly, powder was put in at random and the ball rammed on it without a wad. It was only twenty feet from him, when the man opposite me shot it through the body, which made it fall, and the next bullet finished its career. It was fortunate we were placed in such a position, as there was not so much danger of shooting each other.

This incident taught us a lesson to keep one gun in

reserve. The capture of this bear shewed the superiority of the breech-loader, and that our Minnie rifles were too heavy for such work. We were considered good marksmen, and in our leisure time practised at targets. The use of the rifle is most essential in this country, for none know what moment they may be called on to use their skill.

We had not seen any ships since our arrival ; in fact none but a steamer could have made her way through the intricate passages and amongst such heavy ice. We seldom had our sails set, and were obliged to steam very slowly through it. Although our captain was a thoroughly experienced man, he could not account for the non-appearance of whales. At times we went to the outside in search of them. Very large ones have been caught in the open water in these latitudes. We cruised about for several days without success, and then steamed among the ice off Cape Broer Ruys, but did not see anything alive there except bears.

The days were shortening, which obliged us to find a secure place for the ship at night, and if no large holes of water were found, we made fast to a sconce. One calm clear night, when the harvest moon shone upon the great icefields, the scenery was grand in the extreme. Everything being so peaceful and still, all nature appeared at rest. This was a time for serious reflection and study, nothing could be more sublime. Words fail to express the solemnity of the scene. In our busy life we have seldom time to admire such a sight as this, as our thoughts are otherwise occupied.

The time was drawing on when we might expect a change in the weather. We were about sixty miles from the outside, and still nothing to be seen to make the voyage a payable one. It was considered a waste of time to remain any longer, so we steamed clear of the ice, and made the best of our way home. It was incredible the number of miles we had sailed and steamed, without obtaining what we sought for. After reaching Shetland, we discharged our men as usual, and once more arrived at Dundee, and moored safely in the dock.

CHAPTER XVIII.

THE following year, 1865, I went in the Diana of
Hull, as mate with Captain Gravill, to the seal
fishing at Greenland. This vessel was fitted with moderate
steam power, some iron tanks and many large casks holding
upwards of six hundred gallons each. There was a great
difference between this ship and the powerful steamers
belonging to Dundee.

She had formerly belonged to Bremen, but was sold to
Hull, and strengthened for the Greenland trade, and had
been very successful on former voyages with Captain
Gravill. We left Hull with the crow's nest aloft, which
attracted much attention, and everything went favourably
with us until we got near Shetland, when we encountered a
very heavy gale from the north-west, which lasted two and
a half days. When it subsided, the little steam power we
had was very useful in making our course good, as we had
been blown some distance to leeward. At Lerwick we
shipped most of the men who had been with us before, as
they desired to follow the fortunes of their old master, well
knowing that if a successful voyage could possibly be made
we could accomplish it. We lay a few days as the
weather was very boisterous, and with the first favourable
wind, weighed anchor and sailed for the sealing grounds.

On our way thither, we had several gales, and eventually fell in with the first ice in latitude 69° north, and longitude 3° west. We sailed among streams and patches of pancake ice for many miles, until we came into the latitude of Jan Mayen's Island. There we met with heavier ice, and cruised about to find the track of seals. Several vessels belonging to Norway and some from Scotland now joined us.

One evening a Norwegian sailing ship, called the Eleisa (Captain Foyne, a most fortunate, religious, good old man, respected and beloved by all who knew him), arrived. We were at the edge of the pack, when he manned a boat and came on board. Our captain and he were very old friends, and both had seen many rough days in the Arctic seas. The object of his visit was to hold a consultation, they being the oldest masters in the trade. Captain Foyne stated that he had searched for the seals from lat. 74° N. to lat. 69° N., but no trace of them could be found. By the position of the ice, it was their opinion that the body of seals could not be far away. While he was on board a strong wind began to blow from the eastward, and we were close on the lee ice. Captain Foyne at once took his departure, and the ships worked to windward to gain an offing. The snow fell heavily during the night, but the wind moderated.

At daylight the snow had ceased, but the Eleisa was not in sight. We thought he had run into the pack whilst it was slack, before the sea rose. This proved to be the case, and he at once fell in with the seals and got full. This we heard some months afterwards. We were afraid of getting beset, unless we were sure of doing well, as our orders were, if not successful at the sealing, to proceed to Davis's Straits. It would have been very unfortunate if we had got beset here, and lost the whole Davis's Straits east side whaling.

We sailed sometimes well into the pack, but did not see anything to give us the least encouragement, so we squared our yards to a northerly wind for Shetland, and were not long before we anchored in Lerwick harbour.   The captain had been ill several days before we made the land, and we were glad to get medical advice, as we had no doctor on board.   Five men belonging to Hull wished to leave us, and several others belonging to Shetland.   The Hull men were at once discharged.

We were soon ready for sea, and one morning, when all hands were called, two men were absent.   I went on shore and reported them to the sheriff, who was our agent.   He despatched a constable on horseback to their homes, which was eighteen miles away, and brought them back to the ship.   The difficulty was how to keep them.   The authorities would not detain them on shore until we sailed, so they supplied us with handcuffs, and we kept them in custody on board until we got under weigh.   This was very unpleasant, yet it deterred others from deserting.   Three men were shipped in the place of those who had been discharged.   The two who had deserted, feigned sickness, and a doctor was sent for, who examined them, and found them in perfect health. One of them afterwards told me he had a dream that he would never return.   The Shetlanders, I may state, are very superstitious.   In fact, the whole crew of those ships were more or less so ; for instance, when a vessel has been unlucky for some time, a horse shoe will be taken down and put into the fire, then replaced.   Old horse shoes are nailed principally on the masts in the lower deck as a protection against evil spirits, and a producer of good luck.   It was also considered an ill omen for that day, if a seal appeared with its head out of the water ahead of the boat when on the look-out for whales, but the reverse if it appeared astern.   Or if a fox was shot, and brought on board, it was an ill omen, and many more foolish ideas.

A fine, fair wind springing up, we got under weigh.   All was restored to order, the watch was set, and thus ended our

little annoyance.   Our captain was now in his usual health,
and our thoughts were again turned to the whaling.   The
crew was reduced to fifty men, and if more had been
allowed to leave we should have been short-handed.   One
boat's crew makes a vast difference when fishing.   We had
a fine passage, and sighted a very peculiar rock, called
Rockall, which lies in lat. 57° 36 N. and long. 13° 41 W.

This island or rock is little known, and thousands of
people in this country may be asked where it is, and the
answer will be : " I have never heard of such a place,"
although it is adjacent to the British Isles.   It is a lonely
rock situated in the Atlantic, about 184 miles to the westward of St. Kilda in the Hebrides.

At a distance it has the appearance of a vessel under full
sail, owing to its whitened appearance, which is supposed
to be sea birds' dung.   It has been so little visited that
its formation is indifferently known.   Several fishermen
from Hull have fished on the banks around it, and have
obtained good catches of cod, which attain to a great size,
but, owing to the wild state of the weather, that part of the
sea for fishing is not often frequented.

If the truth was known, I dare say that many a stout
vessel has been lost upon the rock and reef adjacent, and
not a vestige or a soul left to tell the tale.   It is so situated
that vessels in former days bound east from America,
and driven by a continuance of southerly winds, have
frequently seen it in the daytime, but there is nothing to
denote its situation by night.   Thus it was a source of
great anxiety for masters of sailing ships in those days.

I believe that only two or three persons have
ever landed on Rockall, and these have only
stayed for a very short period ; so it appears that
such places near our coast are so little known, yet rocks,
shoals, and small islands, thousands of miles away and
seldom seen, are accurately marked on the charts, and

known to navigators, but this small spot so near to us has been the dread of mariners for many years back in misty weather.   At the same time it is almost an impossibility to erect a lighthouse to warn seamen of its proximity.   A landing cannot be obtained, perhaps not more than once in twelve months, on account of the ever-rolling swell of the Atlantic.

To return to our narrative.   The breeze continued fair. We soon rounded Cape Farewell, and ran up the Straits a short distance from the coast.   The ice was tightly packed on the land on the north part of Disco Bay, so we steamed towards Goodhavn to get news from the natives respecting the movements of the other vessels.   They reported that they had not seen any for several days ; all had gone to the westward.   We wished to get into N.E. Bay as soon as possible, so we steamed towards Waigat Straits, which would save us many miles if we could get through.   The straits divide the island of Disco from the mainland, and coal is found there on the surface, but it will not burn alone, the long exposure to the frost having taken the nature out of it.

When we arrived at the south end of Disco, where the land is very low, nearly the whole length of the Straits can be seen from the mast head.   A narrow floe extended from one side to the other, forming a barrier, so we steamed back and tried the other end of Disco, which is sixty miles long.

By the time we got to Hare Island, the ice had drifted from the land.   Three days had been lost, which was very annoying, for immediately we got into the N.E. Bay, and stood to the westward among the floes, we sighted many whales.   A dense fog suddenly came upon us, but it did not deter us from lowering the boats and pulling in the direction of where we heard them blow.   The fog continued so dense that we could scarcely see a boat's length.   Whales were heard blowing on every side, yet we could not get near them.

A light breeze came at last which cleared the fog away, and we were soon fast to a fine, large whale, which was speedily killed by using the bomb lance. In a short time we had four more killed, and then began to flense. By the time we had finished the wind increased, which broke up the floes, and formed them into a pack. If we had been here four or five days sooner we should have killed as many as the ship would hold, but the Diana with her small steam power could not possibly have forced her way through the floe in the Waigat Straits.

The strong breeze blowing on the pack caused a nasty sea at the ice edge, where we saw many whales disporting themselves as though they knew we could not approach them. We lowered the boats several times, but there was too much wind and sea to contend against, and we were glad to be on board again. If we had fastened to any we might have lost the boat and lines, as the fish would be certain to go under the ice. It was a hundred to one against their keeping in the open water, when they were struck, as they generally take the pack where the boats cannot follow, especially when the wind is on it.

The breeze moderated, but the whales had disappeared. During the time we were killing ours, one of the Dundee steamers came up and also got four. The fishery off Disco was a failure this year, but I feel confident there had been many whales in this neighbourhood on account of there being so much open water. Directly any water makes its appearance in Omenak Fiord, commonly called North-East Bay, the whales are seen there, and at other times in the offing, according to the state of the ice.

We cruised about two or three days, without seeing any, and then proceeded north with light variable winds, until we reached Sanderson's Hope, and then steamed among the Vrow Islands until we came to Horse Head Island, and made fast to the land floe on the south side of it. We were comfortably made fast, when a strong northerly gale set in and blew heavily, but, thanks to the island for its shelter, as it kept the heavy floes from drifting upon us. We were obliged to carry extra warps out their full length, as the floe to which we were made fast kept breaking up in pieces. This lasted twenty-four hours, which made us think there would be a fine opening through Melville Bay.

When we reached the Duck Islands, we found the wind had been too northerly to open the Bay, but had drifted the loose floes along the land ice ; when it became calm the ice did not leave the fast floe for some hours. Immediately it eased, we began to steam into the Bay, and were joined by six or seven steamers, but they were soon out of sight. The wind began to blow from the north-west, and as our ship could not steam head to wind, we sent top-gallant yards down and all hands on the ice to track her into a more sheltered position about half a mile further, but it required all our strength, and full power of steam to reach it, and when we had done so we made fast until the wind abated. The loose floes kept squeezing upon the land floe for some time. A dark sky began to loom in the south, and was quickly followed by sudden gusts of wind from the south-west. We were well protected by a strong point of ice, and we began to saw a dock in the bight, and had no sooner got safely into it, when the gale burst upon us with all its fury. Boats, clothes, and provisions were quickly removed a distance from the ship.

The ice came slowly but surely towards us, and pressed heavily upon the land floe. So long as the broad point held we were safe ; if it gave way nothing could save us. When the ice reached the ship it came to a stand-still, but not before we had a hard squeeze. This gale lasted several hours, but during its continuance the weather was clear.

From the mast-head, we could see the floes coming in

contact with the bergs which were aground, and splitting into pieces. We were very thankful that we had got into such a good position, before the gale commenced, or the Diana would have followed many a noble ship to the bottom of Melville Bay.

After the gale came a calm, and the ice parted, leaving us about one and a quarter miles from the edge. We carried the ice-saws and triangles to the outside and commenced sawing towards the ship, and were occupied in this work twenty hours; then had four hours rest. This went on three days, when we were only half way to the ship. It was a very slow process, yet I never heard a murmur, as everybody knew it was necessary. At the end of the third day, we had been resting an hour, when all hands were called. The ice had parted close to our bows, thereby saving us a great deal of time and labour in getting the ship clear. The sawing had weakened the ice, and a change of current effected the rest. We were now all busy launching the boats, getting our clothes, provisions, etc., on board. In a short time we were steaming out of our prison, but only got two miles away and made fast again. The ice remained stationary to the northward, but was altering very much to the southward.

The weather was fine, clear, and calm. There was a strong mirage in the north-west, which looked to an inexperienced person to be a body of water in that direction, but watching it intently a short time, it would change its appearance to a solid mass of ice. This caused a divided opinion, the captain and I could not trace the least signs of water in the north-west; the other officers said they were certain there was a body of water not far off, but the captain very soon decided which way to go. He was not the man to be persuaded by his officers, unless he was sure they were right; owing to his long experience everybody knew his ideas were the best. This shews how some people

N

would rush headlong into danger without thinking of the consequences of being influenced by others.

Our steam being up, we immediately headed the ship to the southward, and in four hours were in open water off the Duck Islands. We had no sooner set the canvas when a gale sprang up from the S.W., which reduced us to close-reefed topsails for thirty hours. If we had remained a few hours longer in our former position, the consequence might have been serious. Those who had secretly thought we had done wrong by returning south were now convinced of their error.

Meteorology in those regions affords full scope for study, and when carefully observed, will give due notice of such sudden gales. Unfortunately, instruments were not issued at this time, and men generally judged the weather by signs in the sky, sounds in the air, etc., which seldom failed an experienced man. It was only those on whom the great responsibility rested that studied these phenomenal changes of the atmosphere and sky. The majority only thought of their watch being ended and getting below.

The life of a thoughtful officer was far from being a pleasant one. It is true that there were gleams of sunshine in his duties, but very few, and to only a small number is left the responsibility and anxiety of a voyage. If an unprofitable one has been made, the master is liable to be dismissed on his arrival in port. Owners want a return for the great outlay they have had, and therefore try another man, although they know the conduct of the former was exemplary.

When the gale subsided we reached to the westward and found the ice slack, forming a deep bight, and twenty miles from the Duck Islands. We pushed our way through the loose ice for a few miles, and then came to a standstill. The weather was exceedingly fine and clear, but there was not the smallest hole of water to be seen from the crow's

nest. It was one massive, unbroken floe, north, south, and
west.

It was no use staying here, so we immediately turned the
ship round and came into open water again, and steamed
southward along the ice edge, which led us close to
Brown's Island, one of the group called the Vrow
Islands. The late gale had brought the ice down, and it
was upheld by numerous bergs that were grounded in the
vicinity, so we steamed further inshore. A dense fog set
in, which obliged us to make fast to an iceberg close to
Wedge Island, so-called from it resembling a wedge.

Two boats were manned with double crews ; a few stayed
in the boats shooting loom and dovekies, but the others
landed and gathered fifty dozen eider duck and loom eggs.
The loom eggs were obtained from the perpendicular cliffs
forming the south end of the island.

Some Shetlandmen were lowered over the edge of the
cliff with ropes round their waists, and gathered them in
buckets, which were soon filled. This is dangerous work,
but the men were used to it on their own islands. In this
manner we got forty dozen eggs. The island is about a
mile long, and half a mile broad. In four hours we returned
on board, well-loaded. The master gave the greater part
to the crew, reserving a small quantity for the cabin.
Another party made an excursion to some rocks in the
neighbourhood, which were from twenty to thirty feet above
high water-mark, and brought thirty dozen more eider duck
eggs and some birds.

The fog cleared away, and we steamed among the rocks
and islands until we came to Sanderson's Hope, and there
found open water with nothing but bergs to be seen to the
southward. A breeze came from the S.W., when we set
sail and stood to the west ice, and plied along it until we
were off Black Hook. The ice being slack, we pushed
our way through floes and large sconces, and at one time

thought we should be able to get to the fishing grounds which lay between Cape Kater and Home Bay, where many heavy whales have been caught at this time of the year.

We struggled slowly on, many times having all hands to assist in warping, and using every effort to force a passage between the floes in expectation of finding water. We came to a large floe without a crack in it, and made fast. We were now 90 miles from Black Hook, and no water to be seen from the crow's nest. The ship was immovable, and in this position we remained some time, anxiously waiting for the ice slacking to retrace our way to open water.

The weather continued fine and clear, yet the ice did not move, and we drove to the southward at the rate of six miles per day for a period of twelve days. A fresh breeze came from the S.W. with dense fog, causing a commotion among the ice, and with great difficulty and danger we managed to reach the east water again, and made our way further south until we came in the latitude of Cape Dyer, where we once more took the ice, which was much broken up ; but a dense fog again set in for two days, which caused us much trouble in finding the leads in the ice. Sometimes we could get no further, so had to warp and steam in and out, which made us consume more coal than we wished, as so much steaming had made a great inroad into our stock.

At last the weather cleared, and we found ourselves in sight of the west water, a few miles to the southward of Cape Searle, and soon reached it, then went north with a fine wind. The water did not extend many miles from the land, and when we came to Merchant's Bay it was full of ice not broken up. We ran along it as far as Cape Broughton on the north side of the bay.

The captain saw a steamer to the north of us, which caused us all to come on deck, as we had not seen any ship for seven or eight weeks and were full of conjectures concerning the steamer. Some thought she belonged to the fleet which left us in Melville Bay, had got full and laid to, securing boats, etc., for the passage home. Our captain was very dejected, and said he was afraid we had made a mistake by returning from Melville Bay. I felt certain it was an impossibility for us to have got through it with our small steam power. I considered that we had done perfectly right, and that no fault could be attached to us for returning. The appearance of the steamer puzzled us for some time, until we saw two ladies on deck. It then occurred to my mind that it was a new whaling steamer I had heard of, called the Eric, which had been purchased by some enterprising merchants, who had received a grant from the Danish government to work a mine for black lead and other minerals on the east coast of Greenland. The place where they intended to land and erect their houses and plant had not been visited for two hundred years or more.

Eventually, we spoke the steamer, which proved to be the Eric. I was sent to invite the gentlemen and captain on board, who told us there was no possibility of getting near the land on the Greenland coast, so they were trying to get into Cape Hooper harbour. The vessel was well supplied with everything necessary for the expedition. There were some Danish miners on board, also the manager and his wife, the doctor and his wife, and other people connected with them. The captain of the Eric said they had been here twelve days, and the ice was in the same position as when they first came. We were at the floe edge, which formed a deep bight a few miles to the southward of Cape Hooper. It was an unbroken field of ice, and there was no appearance of any water to the northward. Our visitors stayed on board a short time, and gave us the details of the venture. They were much disappointed in not being able to get to their destination on the east coast of Greenland, and now there was not the least prospect of

getting into Cape Hooper harbour.   They were considering
whether to go further south to Exeter Sound, and erect their
houses for the winter.   The steamer was to stay with them
as long as she could, and then return to England.

There was no prospect of getting to the northward and no
sign of whales, so we ran to the southward for Cape Searle
or Malemauk Head.   Nearing the harbour it fell calm, and
we got steam up, but not before it was dark, and the moon
shone bright when we arrived opposite the old Esquimaux
encampment.   The lead was used, and the man said he had
bottom at thirty fathoms.   The land being high we appeared
to be close to the beach, but I knew it was not possible to
be so near with that depth of water.   There are twelve
fathoms a quarter of a mile off, and outside of that there are
ninety fathoms.

No one on board had been here before but myself, and
no doubt they thought I was mistaken.   The anchor was
let go, and ninety fathoms of chain paid out, the watch was
set, and the next morning the boats were sent away to the
outside.

The weather continued calm for three days, and when
the boats were away, a small berg drifted alongside, and on
this account soundings were again taken, and found to be
ninety fathoms.   There is no current in this place, and the
ship had not altered her position.   When the boats returned, we had a hard task to heave the anchor up, this
work occupying five hours.

Having got the anchor, we steamed further inshore, and
brought up in the old anchorage.   If it was not for a bar
on the south side, with only eight feet of water, this would
be a splendid harbour.   The only entrance is from the
north, and when the ice comes in from that direction, it
soon fills up, and remains until a strong southerly wind
drives it out.   Several large streams of ice passed by,
but none entered.

One day when the boats were away, two of us had a lively chase after a bear in the water. There were some large pieces of ice, and I tried to get nearer to the animal by going between two large sconces. They closed upon us, and crushed in both sides of the boat. We were in a dangerous predicament, as we had to jump on the ice and hold the boat to keep it from sinking until two others came to our assistance, and towed us on to the beach close by, and the bear was shot by the other harpooner. The ship was lying on the opposite side of the island, seven or eight miles away, and I, with my boat's crew, endeavoured to cross it in order to get another boat, and bring material to temporarily repair the broken one.

This was the first time anyone had crossed the island, and several bets were taken whether we should succeed or not; but we found it a most difficult task, as it was precipitous and the rocks much broken and rotten, making it dangerous to climb, as there was no firm footing. When half way to the top, we had to take off our boots to get safer foot-hold. When we reached the summit, the boats below looked like midges. A gun was fired to let them know we were safe. The descent on the opposite side was sloping, and easily accomplished. When we sighted the ship, we again fired to attract their attention. A boat came for us, and the men expected we had got a whale at the outside, but of course they were much disappointed.

I reported our accident to the master, who at once despatched us with another boat and materials for bringing back the broken one. It was late in the evening when we returned to the ship, thoroughly tired. No water was seen to the northward from the top of the mountain. This caused us to be on the alert in case we might stay too long, trusting to our paltry steam power.

At daybreak, when all hands were called, the harbour was full of ice. It was not packed very tightly, and a boat could manage to reeve among the pieces, so we manned them, but had not gone more than half a mile when the signal was given to return and prepare to get under weigh. The ice became more packed, and it was with great difficulty we steamed out of the harbour and reached the north side. We began to doubt whether we should be able to clear the land, as the ice tightened visibly, but fortunately for us the land was steep, with the exception of one outlying point, which sailors formerly called the Horse Market. I do not know for what reason that name was given to it, unless it was on account of boats belonging to sailing vessels making it a rendezvous in past years. I have seen as many as sixty there at a time, preparing their coffee, etc. The reader may imagine the commotion and scramble for the boats when a whale made its appearance. Some men would get into the wrong boat, and could not get exchanged until evening.

With the assistance of steam and warps, we managed to pass the point and get into a narrow lane of water very close to the land, and had to brace our yards sharp up for fear they might catch the perpendicular cliff. We cleared the land in safety, and got into more open water, where we were able to set our sails. I believe if we had stayed a few hours longer, we should have been fast for the whole winter.

We were exceedingly fortunate in getting clear, for I do not think a worse place than this could be found to winter in, there being no natives within a hundred miles. We dodged about for a few days, but the ice still drove us south, and at last we ran for Exeter Sound, as there are no harbours from Durban to that place. On our arrival, we anchored in what is called the outer harbour. This anchorage is exposed to easterly winds. When there is no ice to shelter it, a heavy swell rolls in. The inner harbour is well sheltered from the sea, but the distance to pull to the outside is ten or eleven miles. We sent our boats away as usual, and whales were plentiful, but moving rapidly across the bay.

One day the boats separated, three being on the north side and three on the south. One to the southward fastened to a whale, and she was killed before the others came up. We towed her to the ship, and flensed her the next day.

The Eric was already in the inner harbour, and had landed materials for houses and provisions for the wintering party. The manager came on board, and offered to engage any of our men who would like to join them. Three Shetland men volunteered their services. The agreement was signed, and they were landed just before we left. The houses were already built, and everything prepared for the company to winter. The wives of the manager and doctor were also going to stay. It was to some extent hazardous on their part. Very few of the Danish miners would remain, as they considered their contract broken by not being able to land on the east coast of Greenland, and that was why volunteers were wanted from our vessel. The following year all the people were brought home, as the expedition proved a total failure.

We were joined by a Dundee steamer, and a sailing ship called the Windward, belonging to Peterhead. The master of the steamer told us they had the greatest difficulty in getting through Melville Bay, to the west land. They had to go to the northward of Carey's Islands before finding a passage across, and burnt most of their coal, yet had not been successful on the west side. Some had got a few whales in coming to the southward.

Our ideas about the Melville Bay route, therefore, turned out to be correct. He said it was utterly impossible for our vessel to have forced her way amongst the huge floes, and that many times they were at a standstill in forcing

their way through them.   Three ships' boats, eighteen altogether, were daily sent on the south side of the Sound.
One of the Windward's boats got fast to a whale close to
the rocks, another rose near to the second boat.   He fired
his harpoon into her, thinking it was the one already
harpooned.   The whale struck the boat a severe blow with
its tail, capsizing it, and throwing the crew into the water.
The cold was intense that morning, and the poor half-
drowned men were picked up as soon as possible, but were
nearly frozen to death.   One of the steamer's boats conveyed
them on board, making them pull to keep their blood in
circulation.

One poor fellow was so exhausted, that they were afraid
he would succumb before they reached the ship.   They
kept working his limbs and rubbing him, and took off some
of their own clothes to wrap him in.   By such means they
managed to get him on board, and into the hands of the
doctor, who soon brought him round.   Four of the Windward's boats were now entangled with the two fish, which
left only one to lance.   We went to their assistance, and
killed them with the hand lance, reserving the bomb lances
for our own use ; it was not a pleasant job for us.   Sometimes they would be so near to the rocks, that a boat could
not get to work on either side.   It was an interesting sight
for a looker-on.   One whale constantly fought with her head,
raising it twenty feet out of the water, the other lashing her
tail about furiously, making a noise which could be heard
far away.   Considering that a moderate-sized whale's tail is
twenty-two feet from tip to tip, it requires a boat to be
cleverly manœuvred to keep out of its reach.   When in
such shoal water they are exceedingly vicious.

The weather being calm, and the sea smooth, permitted
us to get nearer to them with our lances than we otherwise
could have done had there been any wind or sea.   If such
had been the case, they would have lost both.   We ran

considerable risk in helping them, and got very wet.   One boat received a blow which shook it and broke two oars. We afterwards were informed that the officers of the Windward had received strict orders that morning not to assist anyone, yet they were the first to require assistance. Had we been aware such orders had been issued we should have left them to their own resources.   Such selfish instructions were not becoming from any master.

One evening returning on board, one of our harpooners fired his harpoon into a large walrus.   It was killed with great difficulty.   Our whale lances would not penetrate its hide, and our rifles were of a common make, and when fired, the bullets flattened against its thick skull.   When it was killed we could not haul it upon a piece of ice, which was too high above the water, so we towed it to the beach close by.   In passing a point of land, called the Devil's Point, a sudden change of current caught the boats, whirling them round and forcing us to cut adrift from each other, as we were in danger of being capsized.   This is the only place in the whole country that I am aware of which shows such a tide gate.   At a certain time of the tide the water boils like a whirlpool and runs rapidly to the southward.   Elsewhere there is a steady south-going current close inshore at the principal headlands.

The walrus were numerous in this Sound, and we captured four, but could have killed many more.   It is only when we are returning on board that we attempt to get them, as it would not do to cripple our harpoons whilst on the look-out for whales.   Another time I fired the harpoon at one near the ship.   When it rose again it was further off than the line allowed, so I told another boat to fire into it, thinking I had missed, yet there was a strain on the line, and when we hauled it in we found the harpoon had gone through a young one which had jumped in front of the mother at the moment that I fired, so both were killed, but not before the old one charged the boat and tore the plank nearly out with its tusks. The hide of the former when measured was sixteen feet in length and twelve feet in breadth, and its tusks were black with age. It is seldom a whale harpoon can penetrate their hides, which is more than an inch thick and very tough, unless fired from a gun. We have special harpoons and lances for those animals when we go in search of them.

The Windward left us to go to Niatlik, but we remained in company with the other steamer. The bay ice was making very strong during the nights when it was calm or little wind, and it often prevented us getting near any whales, which were plentiful, but the noise of the ice breaking at the bows of the boats alarmed them.

One night it made stronger than usual, and we thought it was time to leave, as the heavy body of ice from the northward was near us, so we both weighed our anchors and steamed to the outside. As the weather was fine, we stayed a short time, and drifted past Cape Walsingham.

It is singular that no whales have ever been captured off this Cape, or in the vicinity. They are often seen near a group of islands called the Crickerton, which are situated half-way on the east side of Cumberland Gulf. If they were seen going up the Gulf on one side and down the other, then we might say they followed the land round those deep bays. I believe they cross from one headland to another, except late in the season, when they are so plentiful in Cumberland Gulf, where they remain until the ice becomes too thick to allow them breathing space.

Having secured our boats, and made everything ready for sea, we shaped our course for home, and getting into warmer weather the whale lines were washed, which is done by towing them astern when the ship is going through the water. They are then carefully dried, and coiled away. Salt water is better than fresh, as it prevents their getting

mouldy. Great care is taken of everything connected with
the whaling gear, such as guns, harpoons, knives and
lances, etc. These are carefully cleaned and coated with
black lead and paint oil to preserve them from rust during
the winter. The ship is thoroughly scrubbed from the
truck to the water edge when the weather permits, so that
when we return home everything is clean until we commence
discharging. We caught a Norwegian crow about six
hundred miles from the land, and kept it until we reached
Shetland. There we gave it its liberty, which it apparently
did not appreciate, as it came back after two days, when we
were twenty-six miles from Lerwick.

A fine breeze had favoured us so far, when we encountered
a very heavy gale, which forced us to lay the ship to under
bare poles, or only a small tarpaulin in the weather rigging.
In this way we lay thirty hours, and were surprised on the
second morning to see the sea covered with porpoises, as far
as the eye could reach. There must have been thousands
of them. Instead of plunging up and down, as they
usually do, they lay nearly motionless. Captain Gravill
said he had only once seen them like that during a heavy
gale off the Cape of Good Hope.

As soon as the weather moderated, we again set our
canvas and proceeded to Lerwick, where we got a good supply
of fresh provisions. In Baffin's time, about 1612, we are
told that the people of these islands used to barter geese,
fowls, sheep, etc., for old or new clothes in preference to
money, and they continue to do so up to the present time.
Contrary winds detained us three days, at the end of which
time a fair wind sprang up. We then got under weigh, and
arrived at Hull in due course.

In coming through the docks, the rigging was crowded
with boys, the boarding of the sealers and whalers having
been the custom for generations by the boys of Hull.

At one time Hull was the principal port in the kingdom

for the Greenland and Davis's Straits fisheries, which formed
a nursery for seamen.   The year 1869 saw the last of them.
The Greenland Yards, so-called on account of the oil being
boiled and the whalebone cleaned there, gave employment
to many people during the winter.

These places are now converted into warehouses, and the
trade is a thing of the past.

This ends my career of seventeen years in the Arctic
regions.

During the time I was in this trade, I occupied every
position on board ship, from apprentice to master.   I have
in this history of a now departed trade, given the times
when the ships left Hull and other ports, for their adventurous voyages in the Arctic regions, and also the times that
we usually returned.   From this it will be seen that the
crews of these whalers and sealers were at home only during
the winter months, and not, in fact, the whole of these, as
they commenced their voyages usually in February, or early
in March.   The result was in my own case (seeing that I
did not miss a year for the seal or whale fisheries for full
seventeen years) that during this long period I never saw
either blossom or fruit upon the trees, and my eyes and
senses were never blessed with the scent of growing flowers,
the sight of ripening corn, or the subsequent harvest
operations.   On the contrary, my most constant surroundings
during those years were ice, snow, fogs, or the boundless
expanse of ocean.

When the trade was at its height, there were occasionally
fair fortunes made, and the crews at times were well
paid.   The life was frequently one of great privation,
and at all times of deep denial and not a little danger.
There was, however, much of fascination in the pursuit of
seals, whales, and other creatures in the far north, and one
cannot help a feeling of regret that these days have passed
away never to return.

At one time it was almost thought that the world would stand still if the supply of fur of seals, and bone and oil of whales should cease. The supplies did cease, but still the world goes on, and what was half a century ago so highly valued, is now scarcely missed.

Science and nature have ministered to man's necessities, and a far better oil for illuminating purposes than whale oil has been supplied in such abundance that the homes of the poor are supplied with a better, cheaper, and more healthy illuminator than whale oil. Thus we have an assurance that the needs of man will always in some form be supplied by a bountiful Providence.

My story is now done, and I would conclude by expressing the hope that I have afforded entertainment, and, beyond that, instruction to my readers, especially to the young, to whom the Greenland and Davis's Straits Fishery, that was once so important, is now but a name.

APPENDIX.

———••———

HISTORY OF THE BARQUE TRUELOVE.

BUILT for the merchant trade, she was launched at Philadephia, United States, in the year 1764, and has consequently reached the patriarchal age of 109 years. Having proved to be a handy, swift sailing craft, the Truelove was employed by the Americans during their first war with England as a privateer, but, being captured by a British cruiser, was purchased in Hull from the English Government about the year 1780.

She was then employed in the wine trade between Oporto and Hull. The little craft being got up in true man-of-war style, with figure-head and quarter galleries, still carried no less than six guns on a side, and was stoutly manned for defence. Seeing that France was at war with Great Britain, the channel and the coasts of Portugal swarmed with hostile cruisers, and the wine trade was carried on at a great risk. But the Truelove boldly ran her own convoy without waiting, as was then the custom, for an armed consort, and although, on many occasions chased by the French, she always managed to escape.

In the year 1784 we find the good ship transformed into a whaler, being strengthened and fortified to encounter the dangers of the icy north. In this trade she was singularly fortunate, braving numerous perils, to which many of her old consorts succumbed, and many a shipwrecked crew has she brought home safely, whose vessel had been crushed by the relentless ice.

In 1835—the most disastrous year in the records of the Davis's Straits whale fishery—the Truelove formed one of the fleet in Melville Bay, when twenty staunch and true vessels were totally lost, and twelve others were seriously damaged by the ice. Although then exposed to the most imminent danger, the subject of our narrative remained unharmed.

In 1836, she again returned in safety, when several of her consorts were frozen up for the winter and more than half of their crews perished from cold and starvation, while other ships were entirely lost. Captain J. Parker, who had been in command of the old ship on many a trying occasion, used to relate numerous instances of her hair-breadth escapes.

He told how, when exposed to a heavy squeeze among the ice floes, the Truelove would quietly rise up on the surface and rest there until the danger was passed, thus avoiding the fate which many of her sisters suffered. This lucky peculiarity is without doubt due to the remarkable and almost unique model of the old ship. Though antiquated, modern builders might do well to copy, in some points at least. In 1861 we find her again among the floes as tough as ever; and while two ships are wrecked—the Anne brig and barque Commerce—close to her, she was, as usual, squeezed up on the ice on several different occasions, when her crew had the utmost difficulty in getting her afloat again, by sawing and blasting the heavy ice from beneath her.

During another voyage she lay for six whole weeks upon the ice in Melville Bay, and considerable labour was needed then to launch her old frame into its proper element; but that was nothing as long as she had saved herself, and was as sound as before. This wonderful vessel must have made not less than eighty voyages to Greenland and Davis's

Straits, crossing the Atlantic and Polar ocean not less than one hundred and sixty times without a single mishap. Her last voyages as a whaler were made in 1867 and 1868. During the long career of the Truelove in the whaling trade she has brought home not less than between three and four hundred whales.

———

The barque has changed owners many times. In shape the barque is very much like the one in which William Penn arrived at the time he made the treaty with the Indians. The sides batter inwards to the top of the gunwales, and this makes the vessel much broader at the water line than the deck. In nautical language, the sides are known as "tumbling home," because they fall in above the bends. They are directly opposite in construction to those of a ship known as a wall-sided vessel.

———

The Letter of Marque pennant, which had been treasured and carefully kept in a locker in the cabin so many years, was destroyed by someone not knowing its value after she left the whaling trade. The history of it was related to me by Captain Parker when I was an apprentice with him on board of the good old barque. She formerly carried the bust of a man for the figure head, but it was taken off on account of the ice accumulating on it. Her bulwark was called pigsty bulwark, *i.e.*, every other plank out to allow water to run freely off the deck. It was filled up in 1854, which made the deck much warmer.

Her speed has been as high as nine knots, but her usual speed is eight knots per hour.

———

The following is a copy of the ship's papers :—

"Certificate of British Registry.—' This is to certify that in pursuance of an Act in the fourth year of the reign of

King William IV., instituted an Act for the registering of
British vessels,'

"Thomas Ward, merchant, and William Ward, merchant,
both of the Town and Borough of Kingston-upon-Hull,
having made and subscribed the declaration required by the
said Act, and having declared that they are sole Owners in
the proportions specified on the back hereof, of the ship or
vessel called the Truelove, of Hull, which is of the burden
of $296\frac{34}{94}$ tons, and whereof John Parker is Master, and that
the said ship or vessel was built in Philadephia, in North
America, in the year 1764, as appears by the Certificate of
Registry granted at this port the 19th of March, 1831, No.
16, now delivered up and cancelled."

By the ship's papers it appears that James Gleadow, tide
surveyor at the Port of Hull, certified the following as the
dimensions of the barque :—

"One deck, three masts, length from main stem to stern
post, 96 feet ; breadth at the broadest part above the main-
wales, 27 feet $\frac{1}{2}$-inch ; depth of hold, 16 feet, 2 inches ;
square rigged, standing bowsprit, squared sterned, carvel
built, no galleries, no figure head."

A ship called the Harmony, belonging to Hull, commanded by Captain J. Parker, was very much damaged by
the ice. Being full of whales, he had her rounded with
chains, and in that condition brought her safely home,
which caused her to be frequently spoken of as the Old
Harmony in Chains, to distinguish her from another vessel
of the same name.

Dr. Scoresby, in his History of the Northern Whale
Fishery, says, "The first attempt made by the English to
capture the whale, of which we have any account, was
n the year 1594."

The Merchants of Hull, who were ever remarkable for their assiduous and enterprising spirit, fitted out ships for the whale fishery as early as the year 1598, being about half a century after its discovery by Sir Hugh Willoughby. The enterprise was not at first very profitable, and our merchants, after a time, nearly ceased to prosecute it ; but after the passing of the Bounty Act, in 1794, their enterprise again assumed a respectable appearance, which, however, declined after the reduction of the bounty.

From 1772 to 1852, a period of eighty years, 194 ships were fitted out, and sailed from this port to the whale fishery ; out of this number eighty were lost, and six more were taken by our enemies in war time. During these eighty years, Hull ships took 85,644 men, an average of 1070 per year ; and during the same period, brought home 171,907 tons of oil, an average of 88 tons of oil per ship per annum, which sold for £5,158,080, being an average of £64,476 per year. In the year 1820, sixty-two ships were sent out, in 1837 only one.

The trade fluctuated from this period, until it was finally abandoned. The last ship sent out was the Diana, which had been fitted as a steamer, her last voyage being in 1869, on returning from which she was lost off Donna Nook, on the 19th October of that year; since which period the trade has been wholly abandoned, as far as Hull is concerned.

1553—Sir Hugh Willoughby discovered Greenland. Spitzbergen.

1576—Martin Frobisher discovered the Straits named after
him.

1586—Davis discovered the Straits bearing his name.

1593—Whaling first commenced.

1607 and 1608—Hudson's Voyage of discovery towards the
North Pole.

1610—Hudson's last voyage.
1611—Sir T. Button's voyage to the whale fishing.
1616—Eight ships went to Greenland.

LIST OF SHIPS FROM HULL.

1754—Pool.
    Berry.
    York.
    Leviathan.
1755—Ann and Elizabeth.
    Berry.
    Boswell.
    Leviathan.
    Mary Jane.
    Pool.
    York.
1756—The same ships.
1757—Ann and Elizabeth.
    Berry.
    Pool.
    York.
1758—Berry.
    Pool.
1759—Berry.
    Leviathan.
    York.
1760—The same ships.
1761—Berry.
    Leviathan.
1762—The same ships.
1763—No ships.
1764— Do.
1765— Do.
1766—Berry.

1767—Berry.
    British Queen.
1771—Ann and Elizabeth.
    Benjamin.
    Berry.
    British Queen.
    Humber.
    King of Prussia.
    Manchester.

1810—35 ships went from Hull.
1811—44   ditto.
1812—49   ditto.
1814—57   ditto.
1818—63   ditto.
1820—62   ditto.
1821—61   ditto.

In 1868, there were 5 German, 5 Danish, 15 Norwegian, and 22 British vessels, which were in company at West Greenland, and obtained 237,000 seals.

1869—The last was the Diana, which was lost at the entrance to the Humber.

List of the Hull Whaling Fleet in the Year 1811.

SHIP'S NAME.	TON-NAGE.	CAPTAINS.	OWNER.	GUNS
Albion	325	Wheldon ...	Dipkin ...	10
Alfred	305	J. Dick ...	Hall & Co.	2
Augusta	390	Beadling ...	J. Briggs ...	10
Aurora	366	A. Sadler...	J. Gelder ...	10
Bernie	276	Hornby ...	G. Neave...	14
Cato	147	Taylor ...	Tew & Co.	—
Cato	303	Weldon ...	Slater & Co.	6
Duncombe ...	270	Taylor ...	Cooper&Co.	6
Eggington... ...	300	Pinkney ...	J. Egginton	6
Ellison	349	Hawkins ...	ditto.	6
Equestris	370	Young ...	Briggs & Co.	6
Everthorpe ...	349	Trueberry ..	Egginton ...	8
Gardiner	360	Horberry...	ditto.	8
Gilder	360	J. Sadler ...	Gelder & Co	8
Harmony	292	McBride ...	F .Bell ...	—
Ingria	317	Webster ...	Marshall ...	8
John	342	Marshall ...	ditto.	8
Konigsberg ...	254	Kirby ...	Egginton ...	—
Laurel	286	Blenkinsop·	Coates&Co.	4
Leviathan	409	Tather ...	Egginton ...	8
Lord Wellington ...	354	H. Rose ...	W. Bolton..	10
Lynx	337	Briggs ...	J. Dipkin...	—
Manchester ...	285	H. Hunter..	Egginton ...	8
Manchester ...	266	Allen ...	G. Levett...	6
Mary and Elizabeth	313	Wake ...	Bolton ...	8
North Briton ...	250	Jameson ...	Hewetson...	6
Perseverance ...	244	Hunter ...	Briggs & Co	4
Prescott	368	Mitchinson.	Lee ...	—

SHIP'S NAME.	TON-NAGE.	CAPTAINS.	OWNERS.	GUNS
Resolution ...	332	Ezard ...	Edgar & Co.	—
Richard	305	Horberry ...	Smith & Co.	—
Royal George ..	375	Greenshaw..	J. Voase ...	12
Samuel	399	Briggs ...	Hewetson...	8
Sarah & Elizabeth	275	Kinsley ...	J. Dewitt ...	6
Sir Henry Mildmay	150	Edington ...	Captain ...	2
Symmetry	342	Clarke ...	W. Bolton..	10
Thomas	355	Taylor ...	Captain ...	10
Thornton	256	Compton ...	W. Lee ...	6
Trafalgar	330	Wake ...	W. Bolton...	6
Truelove	294	T. Foster...	J. Voase ...	14
Valentine	—	Rose ...	—	—
Venerable	336	Killah ...	W. & T. Hall	2
Walker	335	Sadler ...	Gilder & Co.	—
William	350	Orten ...	Marshall ...	8
Zephyr	242	Bell ...	Gilder & Co.	—

BEAR OR CHERIE ISLAND.

BEAR ISLAND was discovered in 1596 by William Barentz, the Dutch Navigator, whilst on his way to discover the north-east passage to the realms of Cathay or China. The sailors went on shore for a boat load of eggs, and on their return were attacked by a gigantic bear, which proved of a most ferocious disposition. She fought with the sailors while "four glasses ranne out," and swam off with a hatchet which had been driven into her back by a sturdy Dutch tar, and was at last with difficulty killed, when she was found to measure thirteen feet in length. In consequence of this adventure, the land thus discovered received the name of Bear Island.

Bear Island was re-discovered and re christened in 1603.

At that time Hull was one of the most enterprising seaports of England, and a few years previously the first whaling ship despatched from this country had left her harbour for the coast of the Island. In the year specified, the good ship Grace, under the command of Stephen Bennett, was sent forth to make new discoveries, and on the 17th of August came in sight of an island in nearly 75° north lat. It was thereupon named Cherie (or Cherry) Island after Sir Francis Cherie, the patron of the expedition.

In the following year the captain returned in the ship Good Speed, and the wealth of birds and whales led to yearly visits from England, chiefly in the interests of the Muscovite Company, who planted their flag upon the Island and took possession of it in 1610. Not only birds, walrus, and whales were brought from this mystic shore, for, in 1605, thirty tons of lead ore were conveyed to England from a jutting vein that had been discovered at the foot of Mount Misery, a hill that derived its name from the desolate appearance of the island, and the forlorn condition in which Bennett and a boat's crew were placed during one of their early visits. The numerous walrus which used to abound here are now almost extinct. Thousands and thousands have been killed, and the Norwegian and Russian hunters have been compelled to seek "fresh fields and pastures new" in the Kara Sea to the east of Nova Zembla, that "ice cave" which was deemed impenetrable to ships within comparatively recent years.

Sea water freezes at 28° 5′ Fahrenheit above zero.
Fresh ,,   ,,   32°   ,,   ,,   ,,
Mercury   ,,   39° 5′   ,,   below ,,

Formerly it was the custom when relieving the watch on board of sailing ships to sing the following ditty :—

"Larbolins stout, you must turn out,
   And sleep no more within ;
For if you do, we'll cut your clew,
   And let starbolins in,"

and *vice versa.*

Larbolins means larboard watch, now called port.
Starbolins means starboard watch.

Unicorns' horns vary from one to eleven feet in length.

At the present time two beautiful specimens may be seen in the Museum of the Hull Trinity House.

One of them was brought from Greenland by the ship Diana, of Hull, in the year 1798. It measures 8 feet 4 in. in length. The other one was brought from Davis's Straits in the ship Thomas, of Hull—Captain W. Brass—in the year 1817, and measures 8 ft. 2 in. in length.

THERE were many words used in this trade which may be unintelligible to those who have not been to the Arctic regions, and which are not used in these days.

A Fall is shouted when a boat first fastens to a whale, and signifies a jump or leap. (This is an old Dutch word.)

A Nip.—In Arctic parlance a nip is when two floes in motion crush by their opposite edges a vessel unhappily entrapped.

Bay Ice.—Ice newly-formed on the surface of the water. It is then in the first stage of consolidation.

Bight.—An indentation in a floe or pack, varying in size.

Boring.—The operation of forcing the ship through loose or pack ice under a heavy press of sail.

Broach to.—To fly up into the wind ; broadside to the wind.

Centipeding is as follows : When it is calm, the boats on both sides of the ship are lowered to the water edge. They are made fast fore and aft to each other and to the vessel, to keep them steady. The off side oars are pulled, which only employs half the men. This is the custom when all hands have been up a long time and require a little rest. (This is a *very* old custom.)

Clew.—Of a hammock or cot.

Clue.—Of a square sail, either the lower corners, reaching down to where tacks and sheets are made fast to it.

Crow's Nest is a cask fixed at the main top-gallant masthead for the master and officers to keep a look out.

Dodging.—Plying easy, with fore or mainyards aback, taking it easy.

Dyce, and very well dyce.—A term now obsolete—for the helmsman to keep the ship as she goes.

Esquimaux.—A name derived from Esquimantsic, or an eater of raw flesh.

Fetch to.—To reach, or arrive at.

Flensing.—To strip the fat or blubber from a whale.

Floe.—A field of floating ice of any extent as beyond the field of vision.

Flurry.—A convulsive movement of a dying whale.

Fresh way.—Increased speed through the water.

Heavy drift ice.—Dense ice, which has a great depth in the water in proportion to its size.

Hole of Water.—A small space of various dimensions surrounded by ice.

Hove to.—The motion of a ship stopped from going her course.

Hummock.—Protuberant lumps of ice thrown up by some pressure upon a field of ice.

Ice Anchor.—A bar of iron tapering to a point, and bent as a pothook, the point entering the ice and the hawser bent to the shorter hook.

Ice Blink.—A streak of lucid whiteness which appears off the ice in that part of the atmosphere adjoining the horizon.

Icebound.—A vessel so surrounded by ice as to be prevented from proceeding on her journey.

Ice sludge.—Small commuted ice or bay ice broken up by the wind.

Land Floe.—A field of ice of any extent attached to the land.

Land Water.—Water which lies between the ice and the land.

Lane of Water.—A narrow channel between two fields of ice ; any open cracks ; a separation of floe offering navigation.

Laid to.—Same as hove to.

Lead.—A narrow channel of water between the floes or broken ice.

Light ice.—That which has little depth in the water—it is not considered dangerous to shipping.

Making Fast.—To secure the vessel with warps.

Making off.—Cutting the blubber into pieces to pass into the bung-hole of a cask.

Milldolling.—When a ship cannot force its way through young or bay ice a boat is hung by a tackle under the bowsprit with its weight in the water. The crew roll it from side to side, which breaks the ice, allowing the ship to follow. (This is a very old custom.)

Open Ice.—Fragments of ice sufficiently separate to admit a ship forcing or boring through them under sail.

Open Pack.—A body of drift ice, the pieces of which, though very near each other, do not generally touch. It is opposed to close pack.

Pack.—A large collection of broken floe huddled together, but constantly varying its position.

Pancake Ice.—Thin floating rounded spots of snow ice.

Patch.—A smaller collection of broken floe ice drifted from the pack, varying in size.

Plying.—Working to windward ; to beat.

Reaching.—A vessel is said to be on a reach when she is sailing by the wind upon any tack.

Sailing Ice.—A number of loose pieces, at a sufficient distance from each other for a ship to be able to pick her way among them.

Sconce.—A smaller field of ice.

Sound Ice.—Is that which makes in the Sounds and attains to a great thickness, and is said to break up only every seven years.

Spectioneer.—(An old Dutch word.) The chief harpooner who directs the cutting operations of stripping the whale of its blubber, also having charge of the harpoons, lances, knives, etc., belonging to the whaling gear.

Standing in or off.—A movement by which a ship advances towards a certain object, or departs from it.

Stream of Ice.—A collection of pieces of drift ice joining each other in a ridge following in a line of current.

Tack.—A ship is said to be on a tack of the side from which the wind comes.

To Tack.—To go about.

Towing.—The crew towing the ship with one or more boats ahead.

Tracking.—The men on the ice dragging the ship with a line along the floe edge, with canvas belts called tracking belts.

Traveller of Boat's Masts, etc.—A rope grummet, or an iron ring, fitted so as to slip up and down a spar or boat's mast.

Walrus.—Its upper canines are developed into large descending tusks of considerable value as ivory.

Whale.—A general term for various marine animals of the order, *Cetacea*, including the most colossal of all animated beings. From their general form and mode of life, they are frequently confounded with fish, from which, however, they differ essentially in their organization. As they are warm-blooded, ascend to the surface to breathe air, produce their young alive, and suckle them as do the land mammalia. The cetacea are divided into two sections :—1. Those having horny plates, called baleen, or "whalebone," growing from the palate instead of teeth, and including the right whales and rorquals, or finners and hump-backs. 2. Those having true teeth and no whale-bone. To this group belong the sperm-whale and various forms of bottle-noses, black fish, grampusses, narwhals, dolphins, porpoises, etc. To the large species of many of these the term "whale" is often applied.

Whale Boat.—A boat varying from 24 to 26 feet in length, and manned by six men. Some few are longer, and manned by seven men but they are not so handy.